Engineering Mechanics

Engineering Mechanics

Edited by **Derek Pearce**

Willford Press

New York

Published by Willford Press,
118-35 Queens Blvd., Suite 400,
Forest Hills, NY 11375, USA
www.willfordpress.com

Engineering Mechanics
Edited by Derek Pearce

International Standard Book Number: 978-1-68285-218-7 (Hardback)

Printed in China.

Contents

Preface

This book was inspired by the evolution of our times; to answer the curiosity of inquisitive minds. Many developments have occurred across the globe in the recent past which has transformed the progress in the field.

Engineering mechanics is one of the most versatile branches of engineering. The scope of this field is vast as it has multitude of possible practical applications. This book is well equipped to educate the readers and enable them to develop a dynamic problem solving approach. It elucidates the concepts like kinematics, thermodynamics, statics, etc., and innovative researches around prospective developments with respect to engineering mechanics. Enabling the reader to think and apply the prescribed methods in real-life scenarios is the prime motive of the book.

This book was developed from a mere concept to drafts to chapters and finally compiled together as a complete text to benefit the readers across all nations. To ensure the quality of the content we instilled two significant steps in our procedure. The first was to appoint an editorial team that would verify the data and statistics provided in the book and also select the most appropriate and valuable contributions from the plentiful contributions we received from authors worldwide. The next step was to appoint an expert of the topic as the Editor-in-Chief, who would head the project and finally make the necessary amendments and modifications to make the text reader-friendly. I was then commissioned to examine all the material to present the topics in the most comprehensible and productive format.

I would like to take this opportunity to thank all the contributing authors who were supportive enough to contribute their time and knowledge to this project. I also wish to convey my regards to my family who have been extremely supportive during the entire project.

Editor

1

Literature Review on Design, Analysis and Fatigue Life of a Mechanical Spring

Supriya Burgul

[1]*PG Scholar (Mechanical), Vishwakarma Institute of Information Technology , supriyaburgu@gmail.com*

Abstract

In this paper there is reviewed some papers on the design and analysis spring performance and fatigue life prediction of spring. There is also the analysis of failure in spring. The aim of this review paper is to represent a general study on the analysis of spring. Compression springs are commonly used in the I.C. Engine valves,2 wheeler horn & many more and are subjected to number of stress cycles leading to fatigue failure. A lot of research has been done for improving the performance of spring. Now the automobile industry has shown interest in the replacement of steel spring with composite spring. In general, it is found that fiberglass material has better strength characteristic and lighter in weight as compare to steel for spring. We can reduce product development cost and time while improving the safety, comfort, and durability of the vehicles produce. The CAE tool has where much of the design verification is now done using computer simulation rather than physical prototype testing.
Keywords: Spring, finite element analysis, FEM, CAE tool.

1. Introduction

Springs are mainly used in the industry as members absorbing shock energy as well as for restoring the initial position of a part upon displacement for initiating a given function. Compression springs are coil springs that resist a compressive force applied axially. Compression springs may be cylindrical, conical, tapered, concave or convex in shape. Coil compression springs are wound in a helix usually out of round wire. Every two-wheeler has a provision for a sounding a horn to be used while communicating so as to warn the passer-by of the approaching vehicle as well as a signal for maintaining a safe distance or to communicate for any other reasons for safety. It has been reported by the warranty/ maintenance department that frequent complaints are being received over the failures of these springs well within their intended life span. The springs must be designed for reliability. The springs must be designed to withstand the cyclic loading during operation. Therefore in this dissertation work it is proposed to carry out the design and fatigue analysis of compression spring used for Horn in a two-wheeler so as to have better performance in terms of longer life.

2. Some of the important design considerations in spring work

To adhere to proper procedures and design considerations, some of the important design considerations in spring work are outlined here.

2.1. Selection of material for spring construction

a. Space limitations: Do you have adequate space in the mechanism to use economical materials such as oil-tempered ASTM A229 spring wire? If your space is limited by design and you need maximum energy and mass, you should consider using materials such as music wire, ASTM A228 chrome vanadium or chrome silicon steel wire.

b. Economy: Will economical materials such as ASTM A229 wire suffice for the intended application?

c. Corrosion resistance: If the spring is used in a corrosive environment, you may select materials such as 17-7 PH stainless steel or the other stainless steels (301, 302, 303, 304, etc.).

d. Electrical conductivity: If you require the spring to carry electric current, materials such as beryllium copper and phosphorous bronze are available.

e. Temperature range: Whereas low temperatures induced by weather are seldom a consideration, high-temperature applications will call for materials such as 301 and 302 stainless steel, nickel chrome A286, 17-7 PH, Inconel 600, and Inconel X750. Design stresses should be as low as possible for springs designed for use at high operating temperatures.

f. Shock loads, high endurance limit, and high strength: Materials such as music wire, chrome vanadium, chrome silicon, 17-7 stainless steel, and beryllium copper are indicated for these applications.

2.2. General spring design recommendations

a. Try to keep the ends of the spring, where possible, within such standard forms as closed loops, full loops to centre, closed and ground, and open loops.

b. Pitch. Keep the coil pitch constant unless you have a special requirement for a variable pitch spring.

c. Keep the spring index [mean coil diameter, in/wire diameter, in (D/d)] between 6.5 and 10 wherever possible. Stress problems occur when the index is too low, and entanglement and waste of material occur when the index is too high.

d. Do not electroplate the spring unless it is required by the design application. The spring will be subject to hydrogen embrittlement unless it is processed correctly after electroplating.

Hydrogen embrittlement causes abrupt and unexpected spring failures. Plated springs must be baked at a specified temperature for a definite time interval immediately after electroplating to prevent hydrogen embrittlement. For cosmetic and minimal corrosion protection, zinc electroplating is generally used, although other plating such as chromium, cadmium, and tin are also used per the application requirements. Die springs usually come from the die spring manufacturers with coloured enamel paint finishes for identification purposes.

2.3. Special processing either during or after manufacture

a. Shot penning improves surface qualities from the standpoint of reducing stress concentration points on the spring wire material. This process can also improve the endurance limit and maximum allowable stress on the spring.

b. Subjecting the spring to a certain amount of permanent set during manufacture eliminates the set problem of high energy versus mass on springs that have been designed with stresses in excess of the recommended values. This practice is not recommended for springs that are used in critical applications.

2.4. Stress considerations

Design the spring to stay within the allowable stress limit when the spring is fully compressed or "bottomed." This can be done when there is sufficient space available in the mechanism, and economy is not a consideration. When space is not available, design the spring so that its maximum working stress at its maximum working deflection does not exceed 40 to 45 percent of its minimum yield strength for compression and extension springs and 75 percent for torsion springs. Remember that the minimum tensile strength allowable is different for differing wire diameters; higher tensile strengths are indicated for smaller wire diameters.

3. Literature Review

K. Michalczyk [1] The analysis of elastomeric coating influence on dynamic resonant stresses values in spring is presented in this paper. The appropriate equations determining the effectiveness of dynamic stress reduction in resonant conditions as a function of coating parameters were derived. It was proved that rubber coating will not perform in satisfactory manner due to its low modulus of elasticity in shear. It was also demonstrated that about resonance areas of increased stresses are wider and wider along with the successive resonances and achieve significant values even at large distances from the resonance frequencies.

B. Pyttel , I. Brunner, et al. [2] Long-term fatigue tests on shot peened helical compression springs were conducted by means of a special spring fatigue testing machine at 40 Hz. Test springs were made of three different spring materials – oil hardened and tempered SiCr- and SiCrV-alloyed valve spring steel and stainless steel. With a special test strategy in a test run, up to 500 springs with a wire diameter of d = 3.0 mm or 900 springs with d = 1.6 mm were tested simultaneously at different stress levels. Based on fatigue investigations of springs with d = 3.0 mm up to a number of cycles N = 109 an analysis was done after the test was continued to N = 1.5 _ 109 and their results were compared. The influence of different shot peening conditions were investigated in springs with d = 1.6 mm. Fractured test springs were examined under optical microscope, scanning electron microscope (SEM) and by means of metallographic microsections in order to analyse the fracture behaviour and the failure mechanisms. The paper includes a comparison of the results of the different spring sizes, materials, number of cycles and shot peening conditions and outlines further investigations in the VHCF-region.. For comparison the results for the springs with d = 1.6 mm and d = 3.0 mm and Ps = 98% are summarised in Fig. 1. Except for springs made of the stainless steel wire, the fatigue strength of springs with d = 3.0 mm is higher than for springs with d = 1.6 mm. The size effect would imply higher fatigue strength for smaller wire diameters.

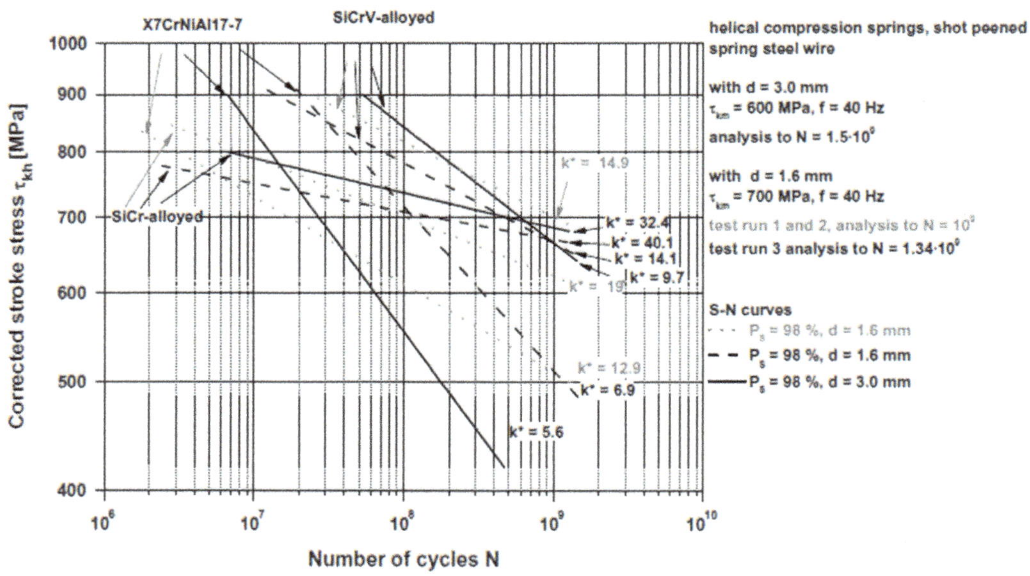

Fig. 1 Comparison of fatigue strength of springs made of various spring steel wires with d = 1.6 mm and d = 3.0 mm at P_s = 98%.

Touhid Zarrin-Ghalami, Ali Fatemi[3] Elastomeric components have wide usage in many industries. The typical service loading for most of these components is variable amplitude and multiaxial. In this study a general methodology for life prediction of elastomeric components under these typical loading conditions was developed and illustrated for a passenger vehicle cradle mount. Crack initiation life prediction was performed using different damage criteria. The methodology was validated with component testing under different loading conditions including constant and variable amplitude in-phase and out-of-phase axial–torsion experiments. The optimum method for crack initiation life prediction for complex multiaxial variable amplitude loading was found to be a critical plane approach based on maximum normal strain plane and damage quantification by cracking energy density on that plane. Rain flow cycle counting method and Miner's linear damage rule were used for predicting fatigue life under variable amplitude loadings. The fracture mechanics approach was used for total fatigue life prediction of the component based on specimen crack growth data and FE simulation results. Total fatigue life prediction results showed good agreement with experiments for all of the loading conditions considered.

Wei Li , Tatsuo Sakai , et al. [4] Very high cycle fatigue (VHCF) properties of a newly developed clean spring steel were experimentally examined under rotating bending and axial loading. As a result, this steel represents the duplex S–N property only for surface-induced failure under rotating bending, whereas it represents the single S–N property for surface-induced failure and interior inhomogeneous microstructure-induced failure under axial loading. Surface small grinding defect-induced failure is the predominant failure mode of this steel in VHCF regime. The surface morphology of the interior inhomogeneous microstructure with distinct plastic deformation is much rougher than that of the ambient matrix, which means the stress concentration resulted from the strain inconsistency between the micro structural in homogeneity as soft phase and the ambient matrix as hard phase plays a key role in causing interior crack initiation. Considering the effect of surface compressive residual stress, the threshold stress intensity factor for surface small defect-induced crack propagation of this steel is evaluated to be 2.04 MPam1/2, which means that the short crack effect plays a key role in causing the surface small defect-induced failure of this steel in the VHCF regime. From the viewpoint of defect distribution, surface and interior failure probabilities are equivalent under a fixed characteristic value of defect density. If the interior defect size is less than or even equal to the surface defect size, surface defect-induced failure will become the predominant failure mode in VHCF regime, especially under rotating bending.

Sid Ali Kaoua a, Kamel Taibi a, et al. [5] This paper presents a 3D geometric modelling of a twin helical spring and its finite element analysis to study the spring mechanical behaviour under tensile axial loading. The spiralled shape graphic design is achieved through the use of Computer Aided Design (CAD) tools, of which a finite element model is generated. Thus, a 3D 18-dof pentaedric elements are employed to discrete the complex ''wired-shape" of the spring, allowing the analysis of the mechanical response of the twin spiralled helical spring under an axial load. The study provides a clear match between the evolution of the theoretical and the numerical tensile and compression normal stresses, being of sinusoidal behaviour. The overall equivalent stress isovalues increases radially from 0_ to 180_, being maximal on the internal radial zone at the section 180_. On the other hand, the minimum stress level is located in the centre of the filament cross section.

B. Pyttel , D. Schwerdt, et al. [6] The paper gives an overview of the present state of research on fatigue strength and failure mechanisms at very high number of cycles (Nf > 107). Testing facilities are listed. A classification of materials with typical S–N curves and influencing factors like notches, residual stresses and environment are given. Different failure mechanisms which occur especially in the VHCF-region like subsurface failure are explained. There micro structural in homogeneities and statistical conditions play an important role. A double S–N curve is suggested to describe fatigue behaviour considering different failure mechanisms. Investigated materials are different metals with body-centred cubic lattice like low- or high-strength steels and quenched and tempered steels but also materials with a face-centred cubic lattice like aluminium alloys and copper. Recommendations for fatigue design of components are given.

Stefanie Stanzl-Tschegg [7] Ever since high-strength steels were found to fail below the traditional fatigue limit when loaded with more than 108 cycles, the investigation of metals' and alloys' very high cycle fatigue properties has received increased attention. A lot of research was invested in developing methods and machinery to reduce testing times. This overview outlines the principles and testing procedures of very high cycle fatigue tests and reports findings in the areas of crack formation, non-propagating small cracks, long crack propagation and thresholds. Furthermore, superimposed and variable amplitude loading as well as frequency effects are reported.

Yuxin Penga, Shilong Wangb, et al. [8] A stranded wire helical spring (SWHS) is a unique cylindrically helical spring, which is reeled by a strand that is formed of 2~16 wires. In this paper, a parametric modelling method and the corresponding 3D model of a closed-end SWHS are presented based on the forming principle of the spring. By utilizing a PC + PLC based model as the motion control system, a prototype machine tool is designed and constructed, which improves the manufacturing of the SWHS. Via the commercial CAD package Pro/Engineering, numerical simulation is carried out to test the validity of the parametric modeling method and the performance of the machine tool. The scheme of the tension control system is analyzed and the control mechanism is set up, which have achieved the constant tension of each wire. A human machine interface is also proposed to achieve the motion control and the tension control. Experimental results show that the tension control system is well-qualified with high control precision.

A.González Rodríguez, J.M. Chacón, et al. [9] An adjustable-stiffness actuator composed of two antagonistic non-linear springs is proposed in this paper. The elastic device consists of two pairs of leaf springs working in bending conditions under large displacements. Owing to this geometric non-linearity, the global stiffness of the actuator can be adjusted by modifying the shape of the leaf springs. A mathematical model has been developed in order to predict the mechanical behaviour of our proposal. The non-linear differential equation derived from the model is solved, obtaining large stiffness variations. A prototype of the actuator was fabricated and tested for different load cases. Experimental results were compared with numerical simulations for model verification, showing excellent agreement for a wide range of work.

Matjazˇ Mršnik, Janko Slavicˇ, et al. [10] The characterization of vibration-fatigue strength is one of the key parts of mechanical design. It is closely related to structural dynamics, which is generally studied in the frequency domain, particularly when working with vibratory loads. Fatigue-life estimation in the frequency domain can therefore prove advantageous with respect to a time-domain estimation, especially when taking into consideration the significant performance gains it offers, regarding numerical computations. Several frequency-domain methods for a vibration-fatigue-life estimation have been developed based on numerically simulated signals. This research focuses on a comparison of different frequency-domain methods with respect to real experiments that are typical in structural dynamics and the automotive industry. The methods researched are: Wirsching–Light, the a0.75 method, Gao–Moan, Dirlik, Zhao–Baker, Tovo–Benasciutti and Petrucci–Zuccarello. The experimental comparison researches the resistance to close-modes, to increased background noise, to the influence of spectral width, and multi-vibration-mode influences. Additionally, typical vibration profiles in the automotive industry are also researched. For the experiment an electro-dynamic shaker with a vibration controller was used. The reference-life estimation is the rainflow- counting method with the Palmgren–Miner summation rule. It was found that the Tovo–Benasciutti method gives the best estimate for the majority of experiments, the only exception being the typical automotive spectra, for which the enhanced Zhao–Baker method is best suited. This research shows that besides the Dirlik approach, the Tovo–Benasciutti and Zhao–Baker methods should be considered as the preferred methods for fatigue analysis in the frequency domain.

Nenad Gubeljaka, Mirco D. Chapettib, et al. [11] High strength steel grade 51CrV4 in thermo-mechanical treated condition is used as bending parabolic spring of heavy vehicles. Several investigations show that fatigue threshold for very high cycle fatigue depends on inclusion's size and material hardness. In order to determine allowed size of inclusions in spring's steel the Murakami's and Chapetti's model have been used. The stress loading limit regarding to inclusion size and applied stress has been determine for loading ratio R=-1 in form of S-N curves. Experimental results and prediction of S-N curve by model for given size of inclusion and R ratio show very good agreement. Pre-stressing and shot-penning causes higher compress stress magnitude and consequently change of loading ratio to more negative value and additionally extended life time of spring.

4. Conclusion

From above papers to conclude that rubber is not suitable material for the coating due to too low value of its modulus of elasticity in shear. Elastomeric coating has a positive impact on the reduction of dynamic stresses in the spring but also contribute to lowering of resonant frequencies. Shorter total life was observed for out-of-phase loading compared to in-phase loading at the same level for both constant and variable amplitude loadings. In finite element analysis of the mechanical behaviour of the twin helical spring under uniaxial tensile load is conducted & findings are compared against those obtained from a theoretical approach based on a transformation of curvilinear coordinates.

The characterization of the fatigue properties of materials and components at very high numbers of cycles necessitates a careful selection of fatigue loading machinery and measuring devices, as well as a diligent application of testing and evaluation procedures.

The present paper proposes a new model of an adjustable-stiffness spring. The proposed device has four leaf springs with nonlinear elastic deformations. The geometry of the leaf spring can be modified by means of an electric motor that adjusts the stiffness of the spring to the desired value. This paper also proposes a mathematical model that allows the leaf springs to be dimensioned for every specific purpose. A prototype of the spring has been built and tested.

References

1. K. Michalczyk:,"Analysis of the influence of elastomeric layer on helical spring stresses in longitudinal resonance vibration conditions,archives of civil and mechanical engineering, 2013".

2. B. Pyttel ,I. Brunner, B. Kaiser, C. Berger, M. Mahendran," Fatigue behaviour of helical compression springs at a very high number of cycles– Investigation of various influences, International Journal of Fatigue,2013

3. Touhid Zarrin-Ghalami, Ali Fatemi,"Multiaxial,fatigue and life prediction of elastomeric components, International Journal of Fatigue,2013".

4. Wei Li , Tatsuo Sakai, Masami Wakitac, Singo Mimura," Influence of microstructure and surface defect on very high cycle fatigue properties of clean spring steel, International Journal of Fatigue, 2013".

5. Sid Ali Kaoua, Kamel Taibia, Nacera Benghanem, Krimo Azouaoui, Mohammed Azzaza,"Numerical modelling of twin helical spring under tensile loading, Applied Mathematical Modelling ,2011".

6. B. Pyttel , D. Schwerdt, C. Berger,"Very high cycle fatigue – Is there a fatigue limit?, International Journal of Fatigue,2011".

7. Stefanie Stanzl-Tschegg,"Very high cycle fatigue measuring techniques, International Journal of Fatigue,2013".

8. Yuxin Penga,b, Shilong Wangb, Jie Zhoub, Song Leib," Structural design, numerical simulation and control system of a machine tool for stranded wire helical springs, Journal of Manufacturing Systems,2012".

9. A.González Rodríguez a, J.M. Chacón b, A. Donoso c, A.G. González Rodríguezd, "Design of an adjustable-stiffness spring: Mathematical modeling and simulation, fabrication and experimental validation,5th Jul 2011."

10. Matjazˇ Mršnik, Janko Slavicˇ, Miha Boltezˇar, "Frequency-domain methods for a vibration-fatigue-life estimation – Application to real data,International Journal of Fatigue 2013".

11. Nenad Gubeljaka, Mirco D. Chapettib, Jožef Predana, Bojan Sencic,"Variation of Fatigue Threshold of Spring Steel with Prestressing, Procedia Engineering ,2011".

A Brief Author Biography

***1st Author Name** –Mrs. Supriya R. Burgul*

Pursuing M.E.Design at Vishwakarma Institute of Information Technology ,Pune,Maharastra,India, Email- supriyaburgu@gmail.com

2

AN AUTOMATIC APPROACH FOR CAN/ PLASTIC BOTTLE CRUSHER MACHINE

Vishal N. Kshirsagar[1], Dr.S.K.Choudhary[2], Prof. A.P. Ninawe[3]

[1] *P.G. Student, Department of Mechanical Engg, KDK College of Engineering, Nagpur, R.T.M. Nagpur University, Maharashtra, India. vishalk031@gmail.com*
[2] *Asst. Professor, Department of Mechanical Engg KDK College of Engg. Nagpur, RTM Nagpur University, Maharashtra, India. skc14763@gmail.com*
[3] *Asst. Professor, Department of Mechanical Engg KDK College of Engg. Nagpur, RTM Nagpur University, Maharashtra, India. free1rhyme@rediffmail.com*
Author Correspondence: Department of Mechanical Engg, KDK College of Engg., Nagpur.
Email Address: - vishalk031@gmail.com

Abstract

In this project I am going to make this fully automatic by using electronics equipment like sensors, microcontroller, IC's etc. along with mechanical components due to which crushed cans or bottles are need not to segregate from bin and also machine will not run idle. After all process has been done, this crusher may help us to understand the fabrication and designing process as well as electronics equipment that involved in this project. After the design has complete, it was transformed to its real product where the design is used for guideline. Crushers are major size reduction equipment used in mechanical and allied industries which crushes different types of soft and hard materials. The can or bottle crusher machine is widely used in beverage industries or in scrap dealers shop to reduce the volume of the cans/bottles. Hence in this design and analysis of various parts are necessary. The Electric Motor and a microcontroller make up the backbone for this project. Overall, this project involves processes like design, fabrication and assembling procedures. This project is mainly about generating a new concept of can/bottle crusher that would make easier to bring anywhere and easier to crush the can or bottle.

Keywords: Slider Crank Mechanism, Beverage Cans or Plastic Bottles, Electric Motor, Sensors, Microcontroller, IC's.

1. Introduction

A can crusher can be defined as "A device used for crushing aluminum cans or plastic bottle for easier storage in recycling bins thereby giving you extra space by flattening of cans". This project consists of designing and fabrication of an automatic can/bottle crusher machine. In order to reduce the waste, we planned to create a can/bottle crusher that will reduce the volume of aluminum cans/plastic bottles by approximately seventy percent by which transportation volume will increase and transportation cost will reduce. Can crushers are primarily used to save space and for recycling. Also the purpose of this project is to understand the fundamental knowledge of design and mechanism by using fulcrum system and a simple mechanism property. A mechanical can/bottle crusher machine is basically one of the most aid able machines. It helps to reduce the pollution of the environment. Thus helps to create a better place to live in, apart from that, this can/bottle crusher can actually be the future mode of recycles apart from the recycle bins. It can be placed everywhere, in the park, houses, restaurants, malls, canteens even in cars. Using a similar type of a design from the diagram below, but with an added bin below the can crusher concept of recycling can be applied. In this project, development of a recycle bin can/bottle crusher so the cans might crush as flat and look as symmetrically as possible and inserted the bin. The designs are an environment friendly and use simple mechanism properties. The design is so done that the knowledge of designing, mechanism and forces. The study of manufacturing was very important in order to carry out this project to ensure that what are needs to do. This project is about designing and fabricating the Recycle Bin Can/Bottle Crusher Machine to help people easy to crush the can as well as bottle and bring anywhere. This project involves the process of designing the different parts of this crusher machine considering forces and ergonomic factor for people to use. After the design has completed, it was transformed to its real product where the design is used for guideline. This project is mainly about generating a new concept of can/bottle crusher that would make easier to bring anywhere and easier to crush the can or bottle. This machine is developed solely for the purpose of recycling as plastic bottles are harmful to environment and aluminum to plants growth. Many people recycles items like paper, glass, and aluminum, While these efforts are a vital part of the process, the true recycling path continues long after recyclables are collected from household bins or community drop-off centers etc. Collecting, processing, manufacturing, and purchasing recycled products create a closed circle or loop that ensures the value of recycling. Recycling is a series of activities that includes the collection of used, reused, or unused items that would otherwise be considered waste, sorting and processing the recyclable products into raw materials, and remanufacturing the recycled raw materials into new products.

2. Concept

This machine is basically works on the principle of Single Slider Crank Mechanism which is the heart of this machine and it converts rotary motion into a reciprocating machine to crush the Cans/Plastic bottles. In this, link 1 is fixed and link 2 which is a crank is rotating about fixed link 1 and converts this rotary motion into the reciprocating motion of slider (corresponds to the link 4) by means of connecting rod which corresponds to the link 3. This is the inversion of single slider crank which is obtained by fixing link 1.

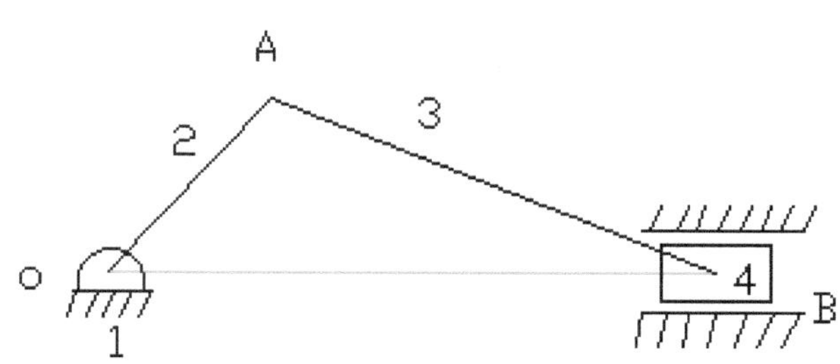

Fig.1- Single Slider Crank Mechanism

3. Working Principle

In this, as the motor rotates the motor is having a smaller pulley connected to its shaft, the smaller pulley through the v-belt is connected to secondary pulley (Larger pulley) on another shaft which rotates it. Another i.e. third pulley (Smaller pulley) is mounted on the same shaft next to the secondary pulley, and is connected to the fourth pulley (Larger pulley) which is on the main shaft by means of v-belt. This arrangement has been done solely for the purpose of reduction of speed. Except that there also automation has been done, such as there I have placed a sensor until and unless sensor senses the obstacle (i.e. a Can/ Plastic bottle) the machine will not start. After machine gets started the power of electric motor is transmitted to the slider mechanism through belt and pulley arrangement and rotary motion of crank is converted into the reciprocating motion of slider thus Cans/Plastic bottles will get crushed. There is a separating bin which is placed beneath a crushing tray to collect the crushed cans/ Plastic bottles also there is a sensor of metal detector type to store the plastic bottles and cans separately.

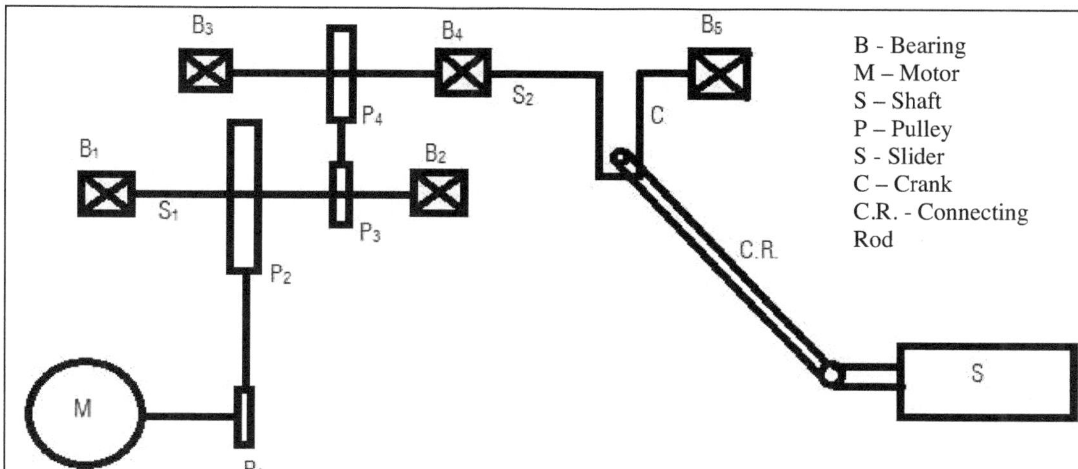

Fig. 2 – Line Diagram of Experimental Set Up

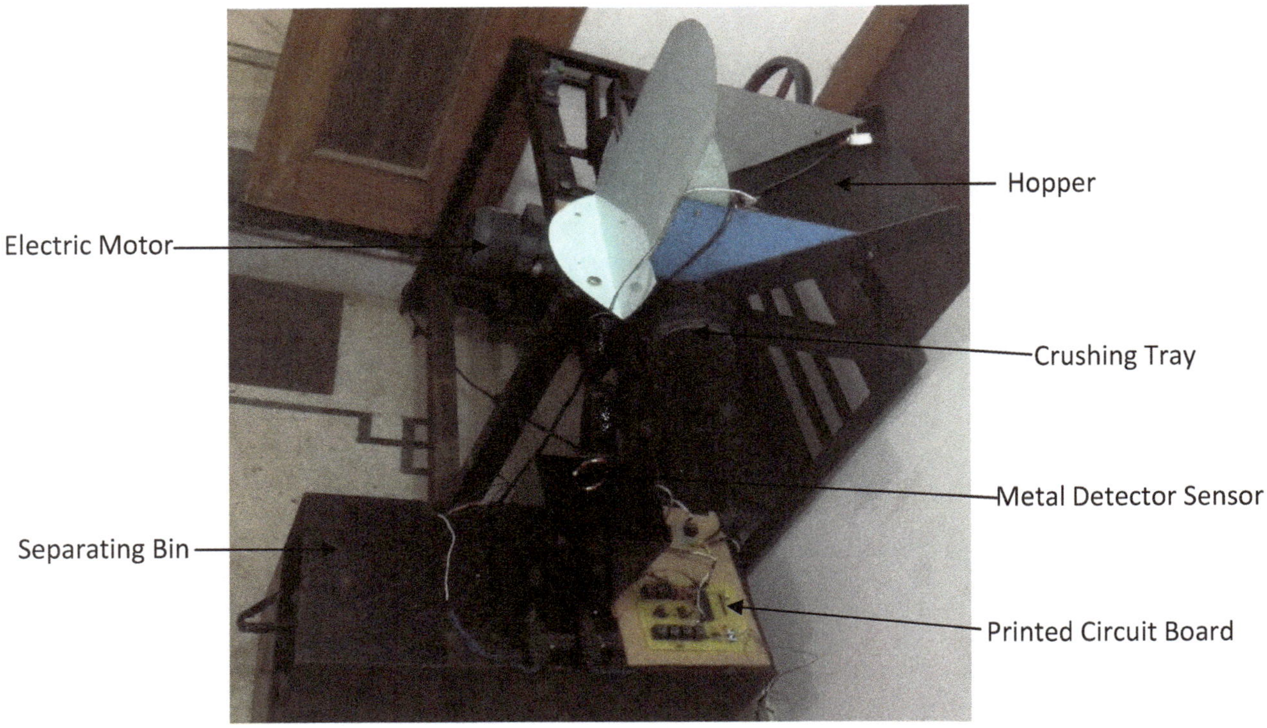

Fig. 3 – Designed Experimental Set Up

The different electronic components which help to make this machine automatic are explained as follows:

1. ATMega16 Microcontroller

The ATmega16 is a low-power CMOS 8-bit microcontroller based on the AVR enhanced RISC architecture. By executing powerful instructions in a single clock cycle, the ATmega16 achieves throughputs approaching 1 MIPS per MHz allowing the system designed to optimize power consumption versus processing speed.

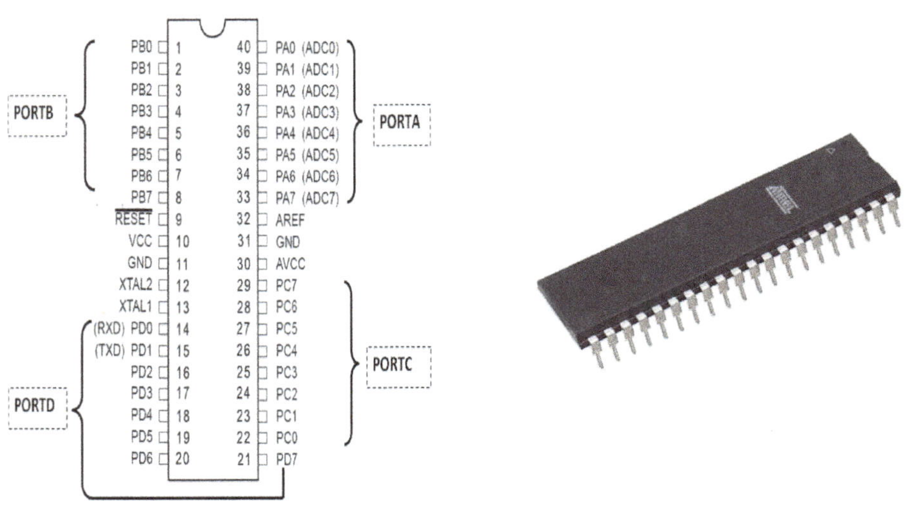

Fig. 4 - ATMega16 Microcontroller

Pin Descriptions

VCC: Digital supply voltage.

GND: Ground.

Port A (PA7..PA0): Port A serves as the analog inputs to the A/D Converter. Port A also serves as an 8-bit bi-directional I/O port, if the A/D Converter is not used. Port pins can provide internal pull-up resistors (selected for each bit). The Port A output buffers have symmetrical drive characteristics with both high sink and source capability. When pins PA0 to PA7 are used as inputs and are externally pulled low, they will source current if the internal pull-up resistors are activated.

Port B (PB7..PB0): Port B is an 8-bit bi-directional I/O port with internal pull-up resistors (selected for each bit). The Port B output buffers have symmetrical drive characteristics with both high sink and source capability. As inputs, Port B pins that are externally pulled low will source current if the pull-up resistors are activated. The Port B pins are tri-stated when a reset condition becomes active, even if the clock is not running.

Port C (PC7..PC0): Port C is an 8-bit bi-directional I/O port with internal pull-up resistors (selected for each bit). The Port C output buffers have symmetrical drive characteristics with both high sink and source capability. As inputs, Port C pins that are externally pulled low will source current if the pull-up resistors are activated. The Port C pins are tri-stated when a reset condition becomes active, even if the clock is not running. If the JTAG interface is enabled, the pull-up resistors on pins PC5(TDI), PC3(TMS) and PC2(TCK) will be activated even if a reset occurs. Port C also serves the functions of the JTAG interface and other special features of the

Port D (PD7..PD0) Port D is an 8-bit bi-directional I/O port with internal pull-up resistors (selected for each bit). The Port D output buffers have symmetrical drive characteristics with both high sink and source capability. As inputs, Port D pins that are externally pulled low will source current if the pull-up resistors are activated. The Port D pins are tri-stated when a reset condition becomes active, even if the clock is not running.

RESET: Reset Input. A low level on this pin for longer than the minimum pulse length will generate a reset, even if the clock is not running. Shorter pulses are not guaranteed to generate a reset.

XTAL1: Input to the inverting Oscillator amplifier and input to the internal clock operating circuit.

XTAL2: Output from the inverting Oscillator amplifier.

AVCC: AVCC is the supply voltage pin for Port A and the A/D Converter. It should be externally connected to VCC, even if the ADC is not used. If the ADC is used, it should be connected to VCC through a low-pass filter.

AREF: AREF is the analog reference pin for the A/D Converter.

2.7805 Voltage Regulator IC

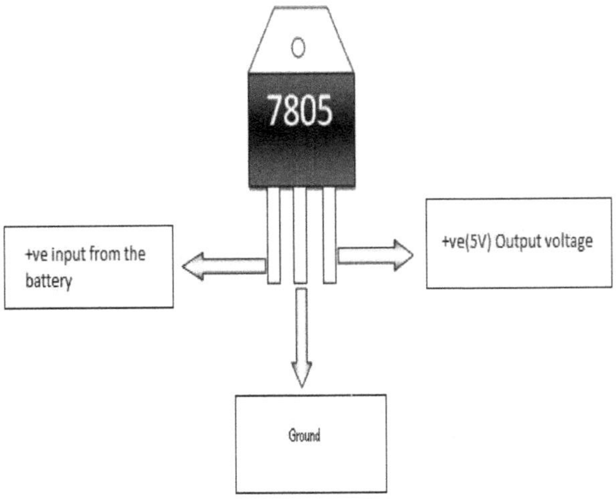

Fig. 5 - 7805 Voltage Regulator IC

7805 is a voltage regulator integrated circuit. It is a member of 78xx series of fixed linear voltage regulator ICs. The voltage source in a circuit may have fluctuations and would not give the fixed voltage output. The voltage regulator IC maintains the output voltage at a constant value. The xx in 78xx indicates the fixed output voltage it is designed to provide. 7805 provides +5V regulated power supply. Capacitors of suitable values can be connected at input and output pins depending upon the respective voltage levels.

Pin Description

Pin No	Function	Name
1	Input voltage (5V-18V)	Input
2	Ground (0V)	Ground
3	Regulated output; 5V (4.8V-5.2V)	Output

3. L293D Motor Driver IC

L293D IC is basically used for driving the inductive loads like DC motors, stepper motors, and relays. It is a 16 pin DIP IC .It will simply amplify the logical input combinations from the microcontroller IC to drive the inductive loads like DC motor in our case. With the help of L293D IC we can drive two motors simultaneously at a time. It has four I/P pins and four O/P pins for controlling the devices by using microcontroller. For a single motor the combinations of two I/P are used for taking the I/P from the microcontroller and after amplification the corresponding two O/P combinations are connected with motor.

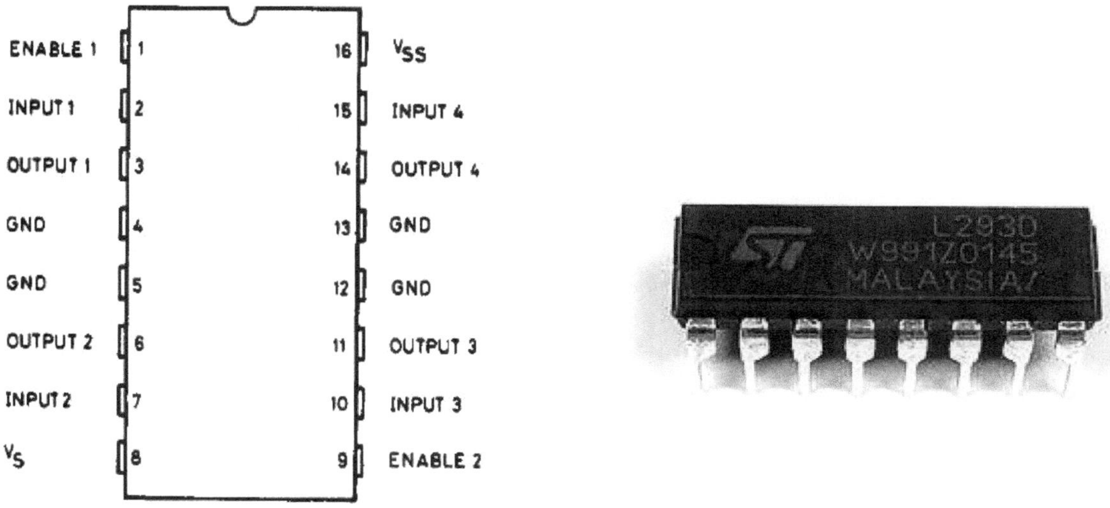

Fig. 6 - L293D Motor Driver IC

4. Metal Detector

This metal detector circuit electronic project is based on transistors and will provide an visual indication using a LED and an acoustic sound that will inform you when a metal is detected. This detector must be powered from a simple 9 volt battery an it needs some little adjustments. To adjust this circuit you need to power it on and move it away from any metal object .You must turn RV2 potentiometer fully clockwise and RV2 potentiometer anti-clockwise. With SW1pressed you must turn RV1 anti-clockwise until the LED is about to light and all further sensitivity adjustments should now be done by turning RV2. To set maximum sensitivity, turn RV2 until the LED is weakly lit. The brightness of the Led indicates the distance and the size of the metal object detected .The L1 and L2 coil must be on the same support that must have a 8mm diameter. L1 has 120 turns from 0.3mm Cu wire and L2 has 43 turns from 0.3mm Cu wire.

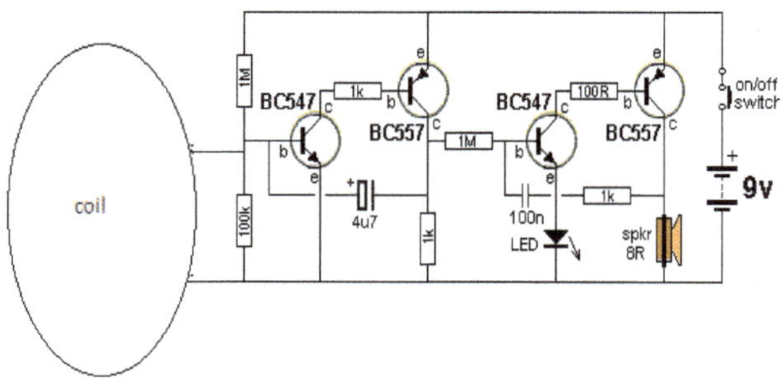

Fig. 7 - Metal Detector Sensor

TSOP Sensor

The TSOP 1738 is a member of IR remote control receiver series. This IR sensor module consists of a PIN diode and a pre amplifier which are embedded into a single package. The output of TSOP is active low and it gives +5V in off state. When IR waves, from a source, with a centre frequency of 38 kHz incident on it, its output goes low. Lights coming from sunlight, fluorescent lamps etc. may cause disturbance to it and result in undesirable output even when the source is not transmitting IR signals. A band pass filter, an integrator stage and an automatic gain control are used to suppress such disturbances. TSOP module has an inbuilt control circuit for amplifying the coded pulses from the IR transmitter. A signal is generated when PIN photodiode receives the signals. This input signal is received by an automatic gain control (AGC). For a range of inputs, the output is fed back to AGC in order to adjust the gain to a suitable level.

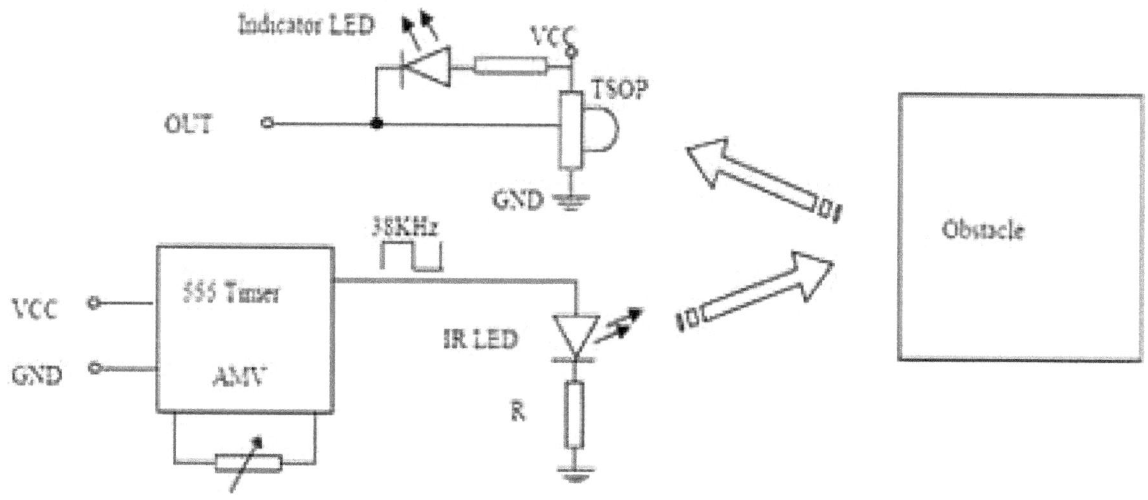

Fig.8 –TSOP Sensor

4. Result and Discussion

As I have taken specimen of 500 ml Plastic bottles as well as Soda Cans having following dimensions

Dimensions of Soda Can-500 ml
Diameter, D = 65 mm
Length, L = 168 mm

Dimensions of Plastic bottle-500 ml
Diameter, D = 58 mm
Length, L = 206 mm
The following tables show the comparison between uncrushed and crushed cans as well as plastic bottles.

Table 1: Reduction in Volume of Cans after Crushing

Sr. No.	Time required	Original Length (mm)	Crushed Length (mm)	Original Volume (mm^3)	Crushed Volume (mm^3)	Reduction in Volume (%)
1.	1.25	168	38	557.47×10^3	126.09×10^3	77.38
2.	1.25	168	37	557.47×10^3	122.77×10^3	77.97
3.	1.3	168	38	557.47×10^3	126.09×10^3	77.38

4.	1.25	168	38	557.47×10^3	126.09×10^3	77.38
5.	1.25	168	37	557.47×10^3	122.77×10^3	77.97
6.	1.3	168	42	557.47×10^3	139.36×10^3	75.00
7.	1.3	168	38	557.47×10^3	126.09×10^3	77.38
8.	1.3	168	38	557.47×10^3	126.09×10^3	77.38

Since, Reduction in Volume in percentage, = $[(V - v)/ V] \times 100$

V – Original Volume
v – Crushed Volume.
Table 2: Reduction in Volume of Plastic bottles after Crushing

Sr. No.	Time required	Original Length (mm)	Crushed Length (mm)	Original Volume (mm^3)	Crushed Volume (mm^3)	Reduction in Volume (%)
1.	1.6	206	58	544.26×10^3	153.24×10^3	71.84
2.	1.65	206	62	544.26×10^3	163.80×10^3	70.00
3.	1.6	206	63	544.26×10^3	166.45×10^3	69.5
4.	1.65	206	63	544.26×10^3	166.45×10^3	69.5
5.	1.7	206	60	544.26×10^3	158.52×10^3	70.87
6.	1.6	206	62	544.26×10^3	163.80×10^3	70.00
7.	1.65	206	60	544.26×10^3	158.52×10^3	70.87
8.	1.65	206	63	544.26×10^3	166.45×10^3	69.5

5. Conclusion

From the above study, it is observed that by arranging different mechanical components along with electronic components machine can be made automatic one to crush the Cans as well as plastic bottles solely to reduce the volume. Thus human efforts can be reduced as well as time also, because sorting system is installed there to separate the bottles and cans to reduce the human fatigue. As above results shows that the volume can be reduced to a large extent i.e. near about seventy percent volume can easily be reduced, thus transportation volume can be increased and transportation cost can be reduced to a large extent and these can be reuse.

References

[1] Mr. Ramkrushna S. More, Sunil J. Rajpal publishes a review paper on "Study of Jaw Plates of Jaw Crusher". International Journal of Modern Engineering Research (IJMER), Vol.3, Issue.1, Jan-Feb. 2013 pp-518-522 ISSN: 2249-6645.

[2] Mr. Shadab Husain, Mohammad Shadab Sheikh presents paper on "Can crusher machine using scotch yoke mechanism". IOSR Journal of Mechanical and Civil Engineering (IOSR-JMCE) e-ISSN: 2278-2064, p-ISSN: 2320-334X PP 60-63.

[3] Mr. Shoichi Kitani, Keiichiro Hayashi, mitsuhiro Yamashina and Keiko Takei Presents a papre on "An Automatic Sorting System For Glass Bottles, Cans and Plastic Bottles".

[4] Mr. Patel Ronak A. presents a paper on "Slider Crank Mechanism for Four bar linkage". IJSRD - International Journal for Scientific Research & Developmentl Vol. 1, Issue 9, 2013 l ISSN (online): 2321-0613.

[5]Cao Jinxi, Qin Zhiyu, Rong Xingfu, Yang Shichun presents a paper on " Experimntal Research on Crushing Force and its Distribution Feature in Jaw Crusher " 2007 Second IEEE Conference on Industrial Electronics and Applications. 1-4244-0737-0/07/$20.00© 2007 IEEE.

[6] Rob Legtenberg, Erwin Berenschot, Miko Elwenspoek and Jan Fluitman presents a paper on "Electrostatic Microactuators With Integrated Gear Linkages For Mechanical Power Transmission". 0-7803-2985-6196 $5.00 © 1996 IEEE.

[7] C. LAROUCI(*), G.FELD (**), JP.DIDIER(*) presents a paper on "Modeling and control of the vehicle transmission chain using electric actuators". 1-4244-0136-4/06/$20.00 © 2006 IEEE

[8] Tiehong GaoJunyi Cao, Dunyu Zhu, Jinzhang Zhi presents a paper on "Study on Kinematics Analysis and Mechanism Realization of a Novel Robot Walking on Water Surface". Proceedings of the 2007 IEEE International Conference on Integration Technology March 20 - 24, 2007, Shenzhen, China. 1-4244-1092-4/07/$25.00 © 2007IEEE.

A Brief Author Biography

Vishal N. Kshirsagar[1] – Presently pursuing M Tech in Mechanical Engineering Design from RTMNU, Nagpur University, Nagpur. Maharashtra, India.

Dr.S.K.Choudhary[2] –Presently working as an Associate Professor in KDK College of Engg., RTMNU, Nagpur University, Nagpur, Maharashtra, India.

Prof. A.P. Ninawe[3] -Presently working as an Assistant Professor in KDK College of Engg., RTMNU, Nagpur University,Nagpur,Maharashtra,India.

DESIGN AND ANALYSIS OF A SINGLE SEATER RACE CAR CHASSIS FRAME

D.LAVANYA[1], G.GURU MAHESH[2], V.AJAY[3] , Dr. C.YUVARAJ[4]

[1]*P.G student, Department of Mechanical Engineering, MITS, Madanapalle.*

[2,3,4] *Department of Mechanical Engineering, MITS, Madanapalle.*

lavanya.mech312@gmail.com[1], gurumaheshg@mits.ac.in[2], ajayv@mits.ac.in[3]

Abstract

This paper describes the design and analyses of a single seater race car chassis frame, according to the specifications of 2014 Formula SAE rule book. The work focuses on the important design parameters around the driver instead of designing the car and the quantitative predictions of different dimensions and mechanical parameters, using the given information and relevant approximations.

Design model was prepared using anthropometric parameters of tallest driver as 95th percentile male was selected to SAE rules book. In this description CATIAV5, HYPERMESH, ANSYS14.5 software's are utilized in order to design and analysis the chassis frame. Based on this design part deformation occurred on the chassis frame at different load conditions and Stress distribution along the each element can be determined.

Keywords- Chassis frame, Deformation, Bending moment, FEA(Finite Element Analysis).

1. Introduction

Formula SAE is an understudy rivalry, supported by the Society of Automotive Engineers (SAE), in which understudies plan, fabricate, and complete with a little equation style race auto. The FSAE rivalry was made to give an instructive experience to school understudies that are practically equivalent to the sort of undertakings they will confront in the work energy. The opposition likewise incorporates some static plan occasions, where the expense and outline of the auto is judged by a board. A chassis is the component in a car that everything else attaches to. The most basic, common chassis design is referred to as the "Ladder Frame" due to its resemblance to a conventional lean-to ladder (Adams, 1992). A "Ladder Frame" consists of two long members that run the length of the automobile and are joined by a set of smaller member perpendicular to the two long members. The other components that make up the vehicle are then mounted to this chassis. In the case of the Ladder Frame, the body and engine are usually mounted to the top of the chassis with the suspension being mounted below. In more recent times the chassis has evolved and in some cars it can be hard to distinguish between what makes up the body of a vehicle and what makes up the chassis. A monologue chassis uses the body as the load carrying component and means that no separate chassis structure is needed. Entire panels carry the load rather than specific members, often these panels are the outermost parts of the body which means that a higher polar moment of Inertia (about the axis running from the front to the rear of the car) is achievable.

2. Chassis Loading

Chassis is a fabricated structural assembly that supports all functional vehicle systems. This assembly may be a single welded structure, multiple welded structures or a combination of composite and welded structures .The first step to designing a chassis is to understand the various loads acting on the structure. The main deformation modes occurring on the chassis is as follows

i. Longitudinal Torsion

ii. Vertical Bending

iii. Lateral Bending

iv. Horizontal Lozenging

i. Longitudinal Torsion

At the point when the connected burdens following up on one or two oppositely contradicted corners of the auto torsion happens. This will encounter while moving an auto on street subjected to compels of distinctive sizes. The edge could be thought as a torsion spring interfacing the two finishes where suspension burdens act. Torsional stacking and undesirable redirections of suspension springs can influence the taking care of and additionally execution of the auto. The imperviousness to tensional misshapening is called solidness spoke to in name. In the execution of a FSAE race auto this is the essential determinant.

ii. Vertical Bending

In this case the frame is act as a simply supported beam and the four wheels as supports and the weight of the driver and components mounted to the frame, such as the engine and other parts are acting at the centre of the frame. Under the effect of gravity produce sag in the frame. Hence the vertical damping force due acceleration or deceleration can increase the vertical deflections.

iii. Lateral Bending

Lateral bending loads are induced in the frame for various reasons, such as road camber, side wind loads and centrifugal forces caused by cornering. The sideways forces will act along the length of the car and will be resisted by the axles, tires and structural members.

iv. Horizontal Lozenging

This deformation is caused when the forward and backward forces applied at opposite wheels. These forces tend to distort the frame into a parallelogram shape. If torsional and vertical bending stiffness are satisfactory, then the structure will generally be satisfactory. The magnitude of these load changes with the operating mode of the car.

3. Load Estimation

As per the standards of the FSAE car parts are designed to withstand 3.5g bump, 1.5g braking and 1.5g lateral forces. The total mass of the car components and driver are listed below:

	Components	Mass (kg)
1	Driver	100
2	Engine	80
3	Drive-train	20
4	Steering	10
5	Battery	04
6	Chassis	30.2
Total		**234.2**

4. Material selection

The chassis material must perform in a high speed environment, with the sharp bend and twist that generate 5-8g forces. In addition to these forces the chassis must be robust and offer some flexibility. Lightweight and stiffness are the most important properties of a chassis and the stiffness of the completed chassis will be affected by the stiffness of the material. After reviewing mechanical properties, availability, cost and other significant factors, following material was selected.

Mechanical Properties of Chassis Material

STEEL GRADE: AISI 4130 steel		
S.No	**Properties**	**Values**
1	Young's modulus	2.05e+011 N/m2
2	Poisson ratio	0.285
3	Density	7850 kg/m3
4	Yield Strength	4.6e+008 N/m2

5. Design requirements (Based on 2014 FSAE Rules Book)

The design of the chassis must work around a number of parameters and constraints in order for it to perform well and for it to be eligible to compete in the competition. These requirements can be broken into several categories which will be discussed below. If any of these requirements are not met, the consequences range from sub-optimal performance to not being eligible to compete in the competition or even chassis failure. So it is clear that all requirements must be carefully considered and even re-visited when designing and building the chassis.

Rules:

The first thing that must be considered when designing the chassis is the 2014 FSAE rules, there is no point in designing a chassis if it will not be allowed to compete in the competition for which it is designed. The FSAE rules require a front and rear roll hoop, a side impact structure, a front bulkhead and supports for the aforementioned components be integrated into the chassis. By representing graphically these 12 requirements one may create a "minimum chassis' Figure 1 which shows the simplest possible configuration of members that include the required components mentioned above. Figure 2 is a side view diagram of what this "minimum chassis" looks like, it does not consider driver ergonomics, cockpit entry or suspension points etc, and is merely a pictorial representation of some of the required members.

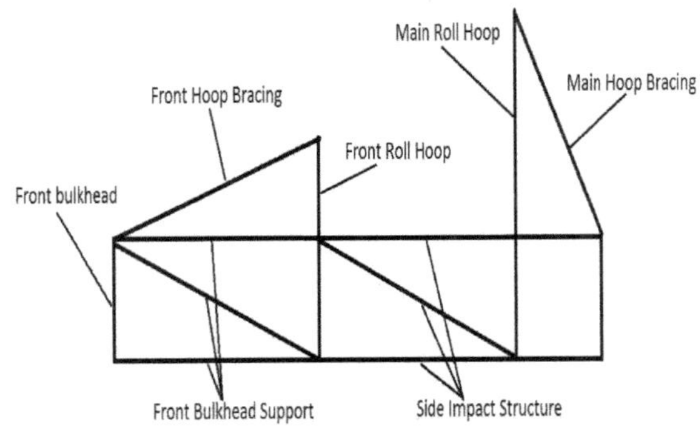

Fig 1. Pictorial representation of the members required by the FSAE rules.

The FSAE rules define a minimum size for all the chassis members shown in Figure 2 and for some other members not shown. To avoid adding un-necessary weight, the chassis design should make best use of the required members so that as few possible additional members are needed. This is where much of the design work needs to be done for the project because as the rules limit many of the members, little design work can be done in optimizing the size of the chassis members.

ITEM or APPLICATION	OUTSIDE DIMENSION X WALL THICKNESS
Main & Front Hoops, Shoulder Harness Mounting Bar	Round 1.0 inch (25.4 mm) x 0.095 inch (2.4 mm) or Round 25.0 mm x 2.50 mm metric
Side Impact Structure, Front Bulkhead, Roll Hoop Bracing, Driver's Restraint Harness Attachment (except as noted above) EV: Accumulator Protection Structure	Round 1.0 inch (25.4 mm) x 0.065 inch (1.65 mm) or Round 25.0 mm x 1.75 mm or Round 25.4 mm x 1.60 mm or Square 1.00 inch x 1.00 inch x 0.049 inch or Square 25.0 mm x 25.0 mm x 1.25 mm metric or Square 26.0 mm x 26.0 mm x 1.2 mm metric
Front Bulkhead Support, Main Hoop Bracing Supports EV: Tractive System Components	Round 1.0 inch (25.4 mm) x 0.049 inch (1.25 mm) or Round 25.0 mm x 1.5 mm or Round 26.0 mm x 1.2 mm "

Fig 2 .2014 Formula SAE Rules for Member size. Adapted from 2014 Formula Student rules (SAE, 2013)

The FSAE rules also require a firewall barrier to isolate the batteries from the driver, it must cover the vertical and horizontal portions of the battery box that face the driver. 2.6mm aluminium sheet is suggested for this firewall but 1mm steel has been approved as an alternative by the FSAE-A rules committee. The chassis must also provide sufficient space for cockpit entry, where the driver enters the cockpit. FSAE rules require a template shown in Figure 4 be able to pass vertically through the cockpit opening until it reaches the height of the top bar in the side impact structure.

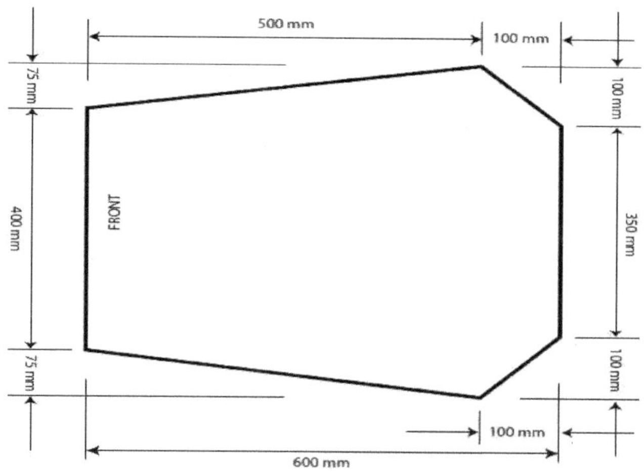

Fig 3: Cockpit entry template

A sufficiently large foot-well area must also be present in the chassis for the driver's legs and feet. The foot-well is the area where the accelerator and brake pedals are located and is where the driver's legs lie when driving. The foot-well area lies between the front suspension pivots so the design will have to work around this area carefully so that the suspension loads are adequately supported, to avoid damaging the chassis and potentially injuring the driver. Fig 4 shows the template that must be able to pass horizontally through the foot-well according to the FSAE rules.

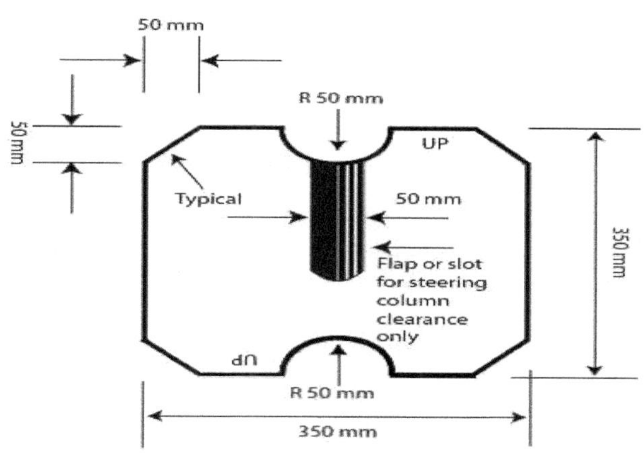

Fig 4: Foot-well clearance template

6. Wire Frame Model

A space frame is designed in several steps that are based on the design considerations. A methodical plan must be followed so that all parameters are considered and the design incorporates every part of the car correctly. The chassis wire frame model is designed by using CATIA V5R20 Generative Shape Design Tool.

<u>*Initial Setup*</u> – First, baseline dimensions like wheelbase, overall length, width, and height were selected. Stemming from these dimensions were roll hoop locations, bulkhead location, engine mounting location, and

wheel centrelines. Once these dimensions were selected, a series of planes were created in CATIA V5 at these points so that these locations could be visualized.

Modelling of Fixed Elements – Fixed elements include roll hoops, front bulkhead, suspension points, and engine mounts. These features will not be moved around during chassis wire frame design iteration so that the number of variables able to be manipulated may be decreased. This allows for a quicker design period so that construction may begin sooner than usual.

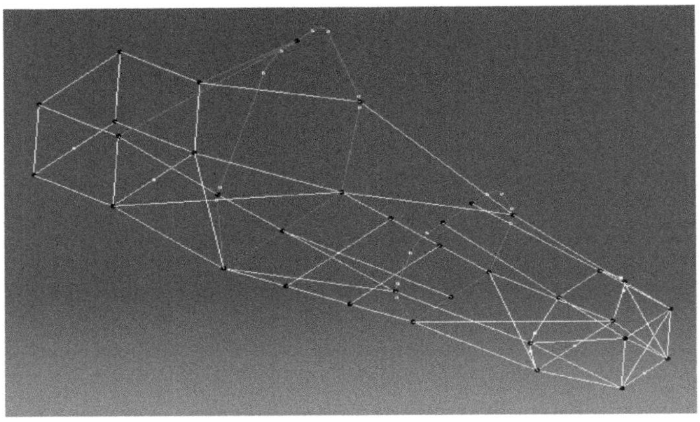

Fig 5: 3D Chassis frame

GENERAL SPECIFICATIONS		
Parameter	**Value**	**Units**
Wheel base	1700	mm
Track width	1120	mm
Ground clearance	160	mm

6.1. Mesh The Wire Frame Model

Altair Hyper Mesh is a high-performance finite element pre-processor that provides a highly interactive and visual environment to analyze product design performance. There are two methods of mesh refinement.

H-refinement- refers to the process of increasing the number of elements used to model a given domain, consequently, reducing individual element size.

p-refinement- element size is unchanged but the order of the polynomials used as interpolation functions is increased.

Steps to be followed to do 1-D mesh

- Import the Chassis wire frame model into the Hypermesh.
- Connect all the lines of the chassis frame.
- Create two components and move the lines into the components by using organise tool bar (because the chassis frame having two variable cross sections).
- create a 1D line mesh (Beam188,element size 10).
- Check the connectivity of the elements.
- Correct the free 1D elements.

- Export the meshed model to the Ansys.

6.2. Finite Element Analysis

The Finite Element Analysis (FEA) is a numerical method for solving problems of manufacturing and mathematical physics. Design geometry is a lot more difficult and the accuracy requirement is a lot higher. We require

– To understand the physical behaviours of a complex entity (strength, heat transfer capability, fluid flow, etc.)
– To predict the performance and behaviour of the design.
– To calculate the safety margin.
– To identify the limitation of the design perfectly.
– To identify the optimal design with confidence.

Steps to be followed to do static analysis

- Import the meshing model from hyper mesh to Ansys.
- The material properties of linear isotropic such as young's modulus= 2.05e5 N/mm^2 and poisons ratio= 0.285 have been defined
- Define cross sections shown in below figs.

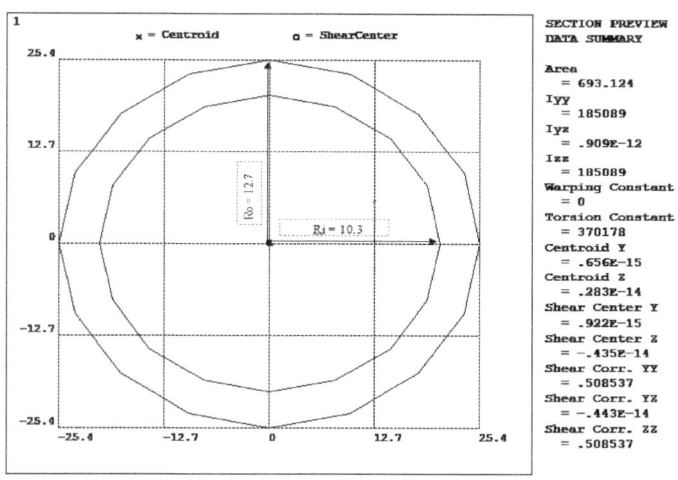

Fig 6: Main and front roll hoop cross sectional area of a chassis frame

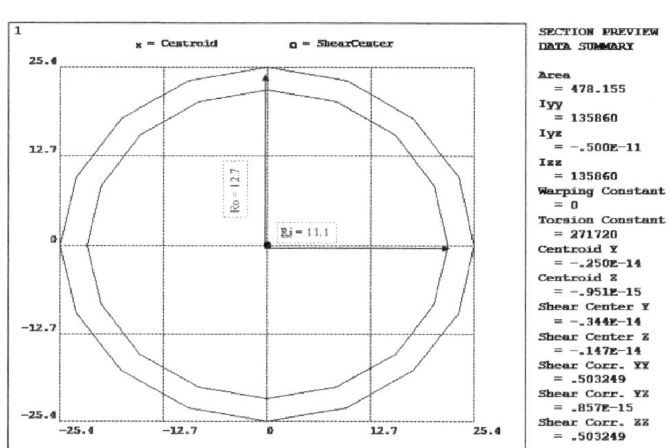

Fig 7: Other sectional members cross sectional area of a chassis frame

- Apply boundary conditions.
- Select type of analysis.
- The problem is solved by using current LS command from the solution menu bar

a. Longitudinal torsion test

Let us consider frame acts like a cantilever beam and its one end is made fixed and other end is subjected to vertical downward force. In this case rear side is fixed and the force is applied at the bulkhead. The force applied is 1480N which is the sum of weight of impact attenuator, driver legs and steering weight.

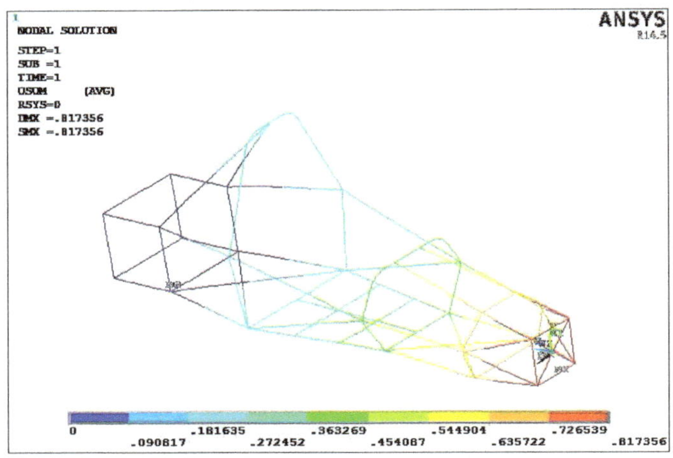

Fig 8: Deformation during longitudinal torsion (load applied at bulk head)

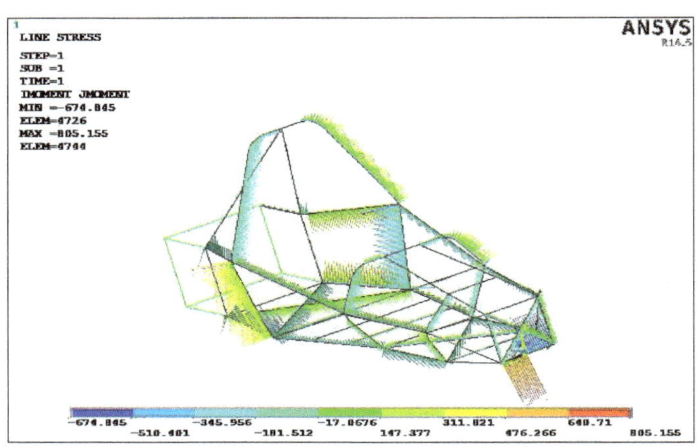

Fig 9: Bending moment during the longitudinal torsion (load applied at bulk head)

In the fig red colour shows the maximum deformed areas and corresponding values is shown in red colour in displacement tree in the bottom. The maximum displacement at the bulk head is 0.81mm and the maximum bending moment 805.15 N-mm and maximum stress occurred in the structure 295.8N/mm^2 .

b. vertical bending

In this case frame is assumed to be in both front and rear suspension is clamped and vertically downward force of 1580N were applied equally in driver cabin.

In the fig red colour shows the maximum deformation and bending moment areas and corresponding values is shown in red colour in stress tree in the left. The maximum deflection and bending moment occurred in the frame is 0.21575mm and 1520.19N-mm respectively. the maximum strain energy capability of chassis frame is 0.143Joules.

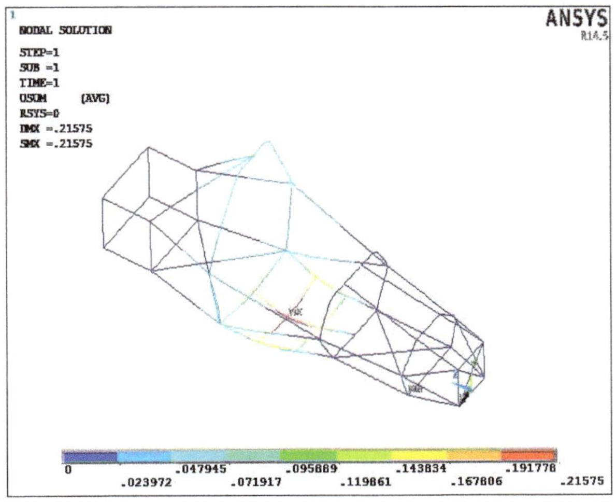

Fig 10: Deformation during vertical bending test (load applied at driver cabin)

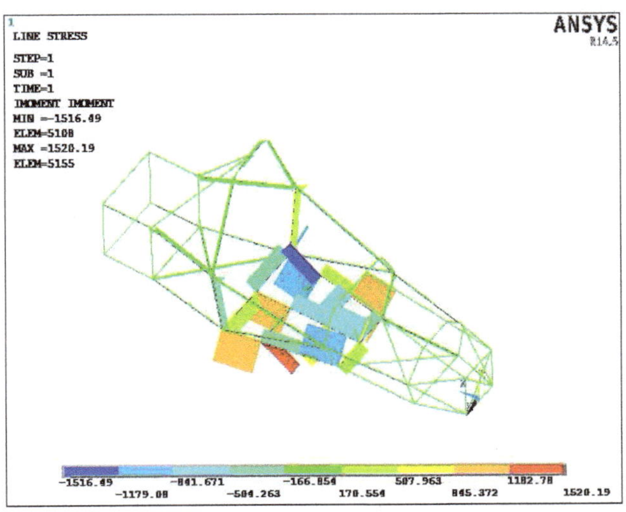

Fig 11: Bending moment during vertical bending test (load applied at driver cabin)

C. Lateral loading

Lateral bending loads are induced in the frame for various reasons, such as road camber, side wind loads and centrifugal forces caused by cornering. the sum of engine and driver mass is act like a lateral cornering force i.e.,2300N applied at side impact bracing of driver cabin.

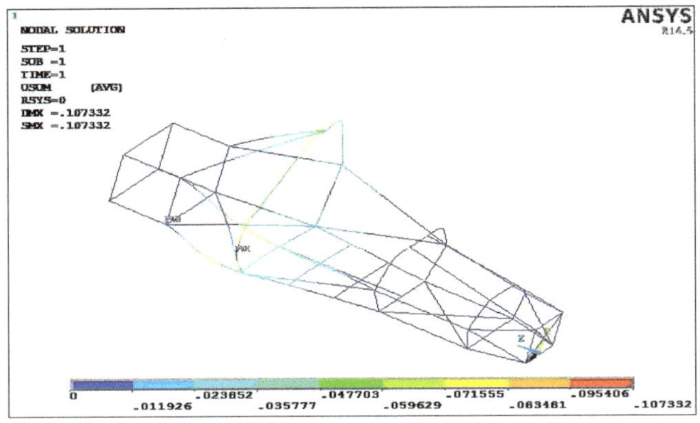

Fig 12: Deformation during lateral bending test

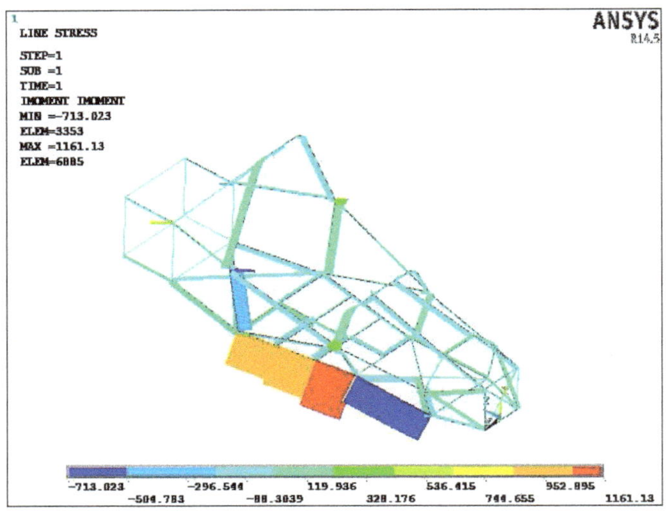

Fig 13: Bending moment during lateral bending test

The maximum deflection 0.107mm and the bending moment 1161 N-mm was observed. The maximum vonmises stress 222.7 N/mm^2 was observed which are in within the permissible limit.

d. Acceleration test

Accelerated testing is an approach for obtaining more information from a given test time than would normally be possible. Generally accelerated testing is in two categories one is acceleration life testing is used for to estimate the life and the other one is acceleration stress testing is used to identify the problem. Due to inertia effect, acceleration forces tend to act in the opposite direction to the motion of the body.

*acceleration of the vechical = force/mass of the vehicle * deltaTime.*

The acceleration force 1150N is applied to the structure in backward direction and the load 1550N was applied downward direction in the driver cabin. The maximum deformation 0.303mm and the bending moment 1692.2N-mm was observed. The resultant von mises stress occurred in the structure is 446.5N/mm^2

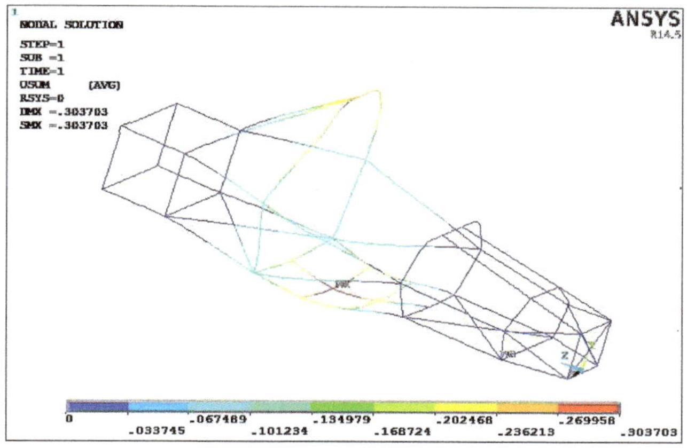

Fig 14: Deformation during Acceleration test (load applied in the driver cabin and apply to the structure in backward direction)

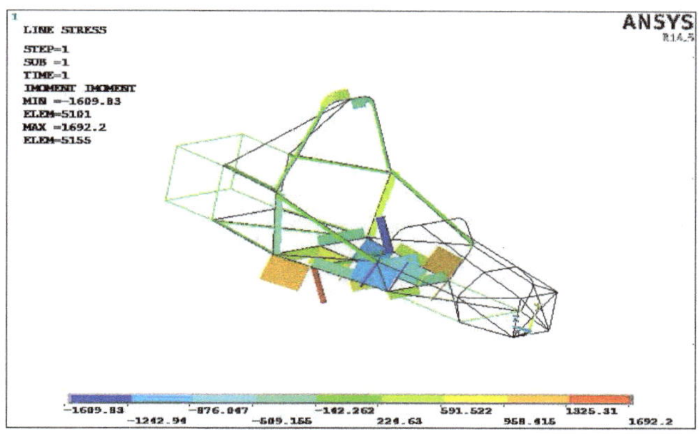

Fig 15: Bending moment during Acceleration test (load applied in the driver cabin and apply to the structure in backward direction)

7. CONCLUSION

S.No	Test	Deflection (mm)	Bending moment (N-mm)	Vonmises stress (N/mm^2)	FOS
1	Longitudinal torsion test	0.81	805.15	295.8	15.55
2	Vertical bending	0.215	1520	267.3	17.2
3	Lateral loading	0.107	1161	222.7	20.6
4	Acceleration test	0.303	1692.2	4465	10.3

In this project I conclude that the chassis frame is with stands for four types of loading conditions and satisfying the FSAE rules and constrains. Thus I would conclude by observing the above results and analyzation that the chassis space frame structure depend on the stiffness and stresses in that particular frame. The stresses obtained above are the well deserved in order to manufacture a chassis space frame as the stresses are very much lower as compared to that of the yield point of the material.

References

1. 2014 Formula *SAE® Rules, SAE International, USA.*

2. M A Abdullah titled with "*Development of formula varsity race car chassis* "*2nd International Conference on Mechanical Engineering Research(ICMER2013) IOP Publishing.*

3. Ravinder Pal Singh titled with " *structural performance analysis of formula sae car*" Jurnal Mekanikal, *December 2010*

4. Mohd Azman. Abdullah titled with " *Design, Analysis and Fabrication of Chassis Frame for UTeM Formula Varsity TM Race Car" International Journal of Mining, Metallurgy & Mechanical Engineering (IJMMME) Volume 1, Issue 1 (2013) ISSN 2320–4060.*

5. Milliken, William F., Milliken, Douglas L., 1997. *Race Car Vehicle Dynamics, Society of Automotive Engineers.*

6. Rouelle, Claude. *Vehicle Dynamics and Race Car Engineering Seminar (2008).*

7. Ryan, A. 2008. *Formula SAE Race Car Analysis: Simulation and Testing of the Engine as a structural member, Retrieved from http://www.fisita.com/students/congress/sc08papers/f2008sc005.pdf on 06 June, 2010 03:40:37 GMT.*

HAVE ANOTHER LOOK ON SERVICE QUALITY IN THE TRACTOR INDUSTRY SERVICE

Dashrath Kumar Patidar[1], Praveen Singh Sisodiya[1] and Prof. Praneeti Saxena[2]

[1]*Research Scholar, Department of Mechanical Engineering, Mahakal Institute of Technology, Ujjain, M.P., India*
Email- Patidarpatel2002@yahoo.co.in,Praveen.mitujjain@gmail.com
[2]*Associate Professor, Department of Mechanical Engineering, Mahakal Institute of Technology, Ujjain, M.P., India*
Email-Praneeti.saxena@rediffmail.com

Abstract

This paper presents results of a investigate designed to weigh up the potential for civilizing productivity and the quality of service for a tractor industries. Tractor is a specialized vehicle used for farming purpose. More than half of the tractors sold in the INDIA are Focus on more than 20 HP and less than 50 HP segment. This compact tractor market has been a growing sector for the last several years. This paper deals with the different issue which is generally important during the selection of tractor. And some another issue related to the service and quality of the tractor. The selection of the tractor done by the quality and what service is provided by the company.

Keywords: Service, Quality, Gaps, SERVQUAL, Customer, Expectations, Perceptions, etc.

---

1. INTRODUCTION

Now more than ever, farmers need to control their tractors efficiently because machinery's carve up of the total cost of crop production is exceeded only by land speculation. Farm managers can increase profits by selecting the right harmonize of equipment for their farm size, labor and tillage program. In many cases, economic survival of a farm operation depends on the farmer's ability to manage and select machinery efficiently.

Well-timed information and performance data should be examined by the farmer during the resolution process so the tractor selected will meet the farm's economic and power requirements. Unfortunately, often a assessment will be made without this information or performance data because the source is unknown or unavailable to the farmer. Knowing what sources of information are typically used by farmers, and which are most beneficial, will enable the specialists, engineers, sales personnel, and educators to provide current and timely information in these sources so the machines selected are the most efficient complement for each individual farm [1].

2. LITERATURE REVIEW

2.1 QUALITY

According to the Oxford Dictionary for the business world, Quality is defined as the degree of excellence [2].
Quality guru J.M.Juran defined quality as 'Fitness for the purpose' [2].
Quality guru Philip Crosby defines quality as 'Conformation to specifications' [2].

2.1.1 Various dimensions of quality

Eight dimensions of product quality management can be used at a strategic level to analyze quality characteristics. The concept was defined by David Garvin. Some of the dimensions are mutually reinforcing, whereas others are not—improvement in one may be at the expense of others. Understanding the trade-offs desired by customers among these dimensions can help build a competitive advantage. Garvin's eight dimensions can be summarized as follows: [3].

A. **Performance**: Performance refers to a product's primary operating characteristics. This dimension of quality involves measurable attributes; brands can usually be ranked objectively on individual aspects of performance.
B. **Features**: Features are additional characteristics that enhance the appeal of the product or service to the user.
C. **Reliability**: Reliability is the likelihood that a product will not fail within a specific time period. This is a key element for users who need the product to work without fail.
D. **Conformance**: Conformance is the precision with which the product or service meets the specified standards.
E. **Durability**: Durability measures the length of a product's life. When the product can be repaired, estimating durability is more complicated. The item will be used until it is no longer economical to operate it. This happens when the repair rate and the associated costs increase significantly.
F. **Serviceability**: Serviceability is the speed with which the product can be put into service when it breaks down, as well as the competence and the behavior of the serviceperson.
G. **Aesthetics**: Aesthetics is the subjective dimension indicating the kind of response a user has to a product. It represents the individual's personal preference.
H. **Perceived Quality**: Perceived Quality is the quality attributed to a good or service based on indirect measures.

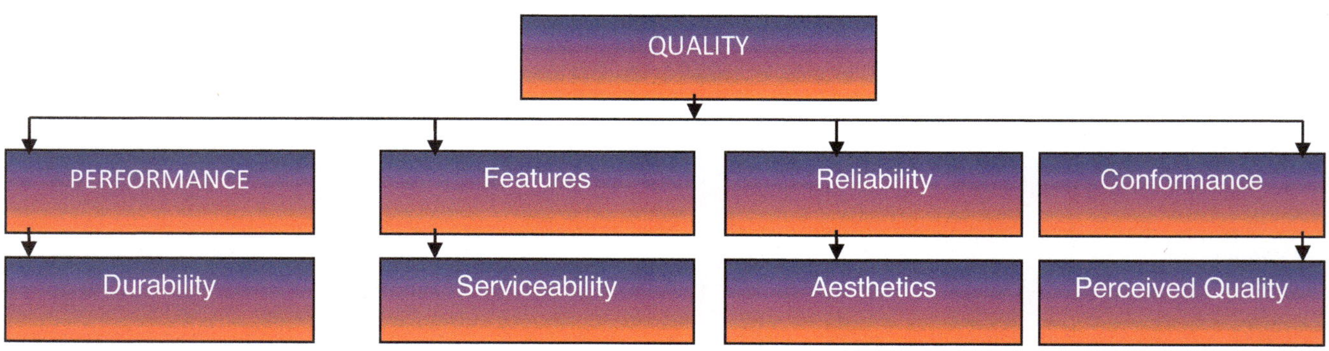

Figure No.1 various dimensions of quality

2.2 SERVICE QUALITY

Service quality is a concept that has aroused considerable interest and debate in the research literature because of the difficulties in both defining it and measuring it with no overall consensus emerging on either [4]. There are a number of different "definitions" as to what is meant by service quality. One that is commonly used defines service quality as the extent to which a service meets customers' needs or expectations [5]. Service quality can thus be defined as the difference between customer expectations of service and perceived service. If expectations are greater than performance, then perceived quality is less than satisfactory and hence customer dissatisfaction occurs [6].

Always there exists an important question: why should service quality be measured? Measurement allows for comparison before and after changes, for the location of quality related problems and for the establishment of clear standards for service delivery. State that, in their experience, the starting point in developing quality in services is analysis and measurement.

Today service industries dominate our economy. The service sector accounts for more than 70% of jobs, and the number continues to grow. Quality improvement looms large in the ongoing success of this sector of our economy. The statistical methods and procedures we examined in preceding chapters are also used for service operations. However major differences exist in the quality characteristics of manufacturing and service [7]. Accordingly, both the measurement process and managements focus different. In service industries not only the products meet the functional requirements of the customer, but the employee behavior must also meet high standards. The total service concept is a combination of technical and human behavioral aspects, and latter are much more difficult to quantify, measure, and control.

Thus service quality means the satisfaction of customers about the service which he is getting. The quality of a service can be broken down into two categories: effectiveness and efficiency. Effectiveness deals with meeting the desirable service attributes that are expected by the customer. For example, the doctor and available facilities in a hospital room, the quality and quantity of food served in a restaurant and the types of checking and savings account available in a bank are related to service effectiveness. Efficiency on the other hand concerns the time required for the service to be rendered.

2.2.1 Service quality characteristics

The quality characteristics are grouped into four categories although exceptions to this group exist; the categories generally summarize factors common to service function industries [8].

A. Human factors and behavioral characteristics:

Service quality is influenced by the attitude and behavior of the provider. Since providers and customers are part of the product, the service their behavior affects the quality of the service. Human factors thus include intensity, eagerness to help, thoughtfulness, complacency, courtesy and so on. Some of these traits can be developed through adequate training some are inherent in the individual. Proper screening of employees and appropriate job assignment are ways to achieve desirable quality characteristics. A primary source of customer's complaint is discourteous behavior.

B. Timeliness characteristics:

A service that is not used in a given span of time cannot be stored for later use. A hospital with empty beds during certain days of a month cannot save them for use in the following month. Thus the timeliness with which a service is performed is critical to customer satisfaction. How long did the customer have to wait in line to cash a check? Characteristics related to timeliness are categorized by the service phase with which they are associated. Categories might include the time to order the service, the waiting time before the service is performed, the time to serve, and the post service time. These characteristics are much more amenable to measurement than behavioral characteristics.

C. Service non conformity characteristics:

These characteristics deal with deviation from target performance levels; a non conformity is a deviation from the ideal level. Examples of such characteristics include the number of errors in processing per 100 vouchers by bank employees, the number of data entry errors per 1000 keystrokes by a data entry operator, the number of billing errors per 100 accounts by a utility company, the number of complaints per 100 guests in a hotel, and so on. The target performance level for these examples is zero nonconformity. The goal of the service organization is to achieve the target level, thus meeting customer expectations, and then exceed it through quality improvement measures. Quality characteristics in this category are well defined and are more readily measured than behavioral characteristics.

D. Facility related characteristics:

The physical characteristics of the facilities associated with a service and its delivery can impact customer satisfaction. The décor of a restaurant, the waiting area in a bus station, and the availabilities of such amenities as a swimming pool or spa in a hotel are examples of quality characteristics for physical facilities that are involved in providing a service. The appearances of a waiter or waitress, a bank teller or an insurance agent are attributes of individuals performing a service. These characteristics are not as clearly defined and measurable as service nonconformities characteristics. They are, however more quantifiable than behavioral characteristics.

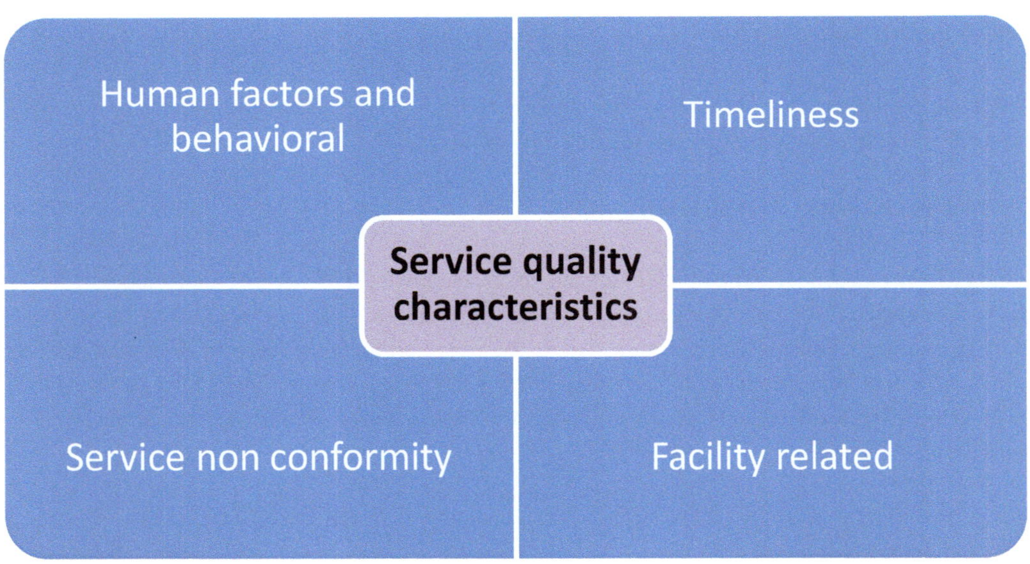

Figure No.2 Service quality characteristics

2.2.2 Service quality dimensions

Service quality has been the subject of considerable interest by both practitioners and researchers in recent years. Definitions of service quality hold that this is the result of the comparison that customers make between their expectations about a service and their perception of the way the service has been performed [9][10].

Several studies have been conducted to identify traditional service quality dimensions that contribute most significantly to relevant quality assessments in the traditional service environment [10]. Identification of the determinants of service quality is necessary in order to be able to specify measure, control and improve customer perceived service quality [11].

Parasuraman et al. identified 10 detailed determinant of service quality through focus group studies: Tangibles, reliability, responsiveness, communication, access, competence, courtesy, credibility, security, and understanding/Knowledge of customer.

Later these ten dimensions were further purified and developed five dimensions-tangibles, reliability, responsiveness, assurance and empathy to measure service quality, SERVQUAL [10].

Tangibles refer to physical facilities, equipment, and appearance of personnel. Reliability means ability to perform the promised service dependency and accurately. Responsiveness means willingness to help customers and provide prompt service. Assurance indicates knowledge and courtesy of employees and their ability to inspire trust confidence. Empathy refers to caring, individualized attention the firm provides its customers.

Walker suggested that the key determinants are product reliability, a quality environment and delivery system that work together with good personal service-staff attitude, knowledge and skills.[12] postulated six criteria of perceived good service quality: professionalism and skills; attitudes and behavior; accessibility and flexibility; reliability and trustworthiness; recovery; reputation and credibility and flexibility; reliability and dimensions attentiveness/helpfulness, responsiveness, care, availability, reliability, access, flexibility,
aesthetics, cleanliness/tidiness, comfort and security.

From the focus group interviews, [13] identified ten determinants of service quality. Virtually all comments consumers made in these interviews about service expectations, Priorities and experiences fall into one of these ten categories. These are reliability, responsiveness, competence, access, courtesy, communication, credibility, security, understanding and tangibles.

Reliability involves consistency of performance and dependability. It means that the firm performs the service right the first time. It also means the firm honors its promises. Specially it involves: accuracy in billing, keeping record correctly, performing the service at the designated time. Responsiveness concerns the willingness or readiness of employees to provide service. It involves timeliness of services that means mailing a transaction slip immediately, calling the customer back quickly and giving prompt service. Competence means possession of the required skills and knowledge to perform the services. It involves: knowledge and skill of the contact personnel, knowledge and skill of operational support personnel, research capability of the organization. Access involves approach, ability and ease of contact. It means: the service is easily accessible by telephone,
waiting time to receive service is not extensive, hours of operation are convenient and location of service facility is convenient. Courtesy involves politeness, respect, consideration, and friendliness of contact personnel. It includes-consideration for the consumer's property, clean and neat appearance of public contact personnel. (ibid) Communication means keeping customers informed in language they can understand .It also means listening to customers. It may mean that the company has to adjust its language for different consumers-increasing the level of sophistication with a well-educated customers and speaking simply and plainly with a novice. It involves: explaining the service itself, explaining how much the service will cost, assuring the customer that a problem will be handled. Credibility involves trusts worthiness, believability, honesty; it involves having the customer's best interests at heart. Contributing to credibility is: company reputation, personal characteristics of the contact personnel. The degree of

hard sell involved in interaction with the customer. Security is the freedom from danger, risk or doubt. It involves: physical safety, financial security and confidentiality. Understanding the customer means making the effort to understand the customer's need. It includes: learning the customer's specific requirements, providing individualized attention, recognizing the regular custom. Tangibles includes the physical evidence of the service, physical representations of the service, such as a plastic credit card or bank statement, other customers in the service facilities. [14] developed an application for measuring retail banking service quality, which consists of 28 attributes including four service quality dimensions such as: accessibility; competence; accuracy and friendliness; and tangibles. The accuracy and friendliness dimension turned out to be the most important factor out of four determining banking preference, followed by competence, tangibles, and accessibility. [15] Proposed an alternative measure of perceived service quality in retail banking that comprises 31 items with six underlying key dimensions. These dimensions are: effectiveness and assurance, access, price, tangibles, service portfolio and reliability.

2.2.2 Five service quality characteristics that please any customers. Discover desired qualities of service providers

A. **Responsiveness:** willingness to help and provide services
B. **Assurance:** knowledge and courtesy of staff
C. **Empathy:** caring and individualized attention
D. **Tangibles:** availability of equipment and appearance of physical environment
E. **Reliability:** ability to perform the promised service dependably and accurately

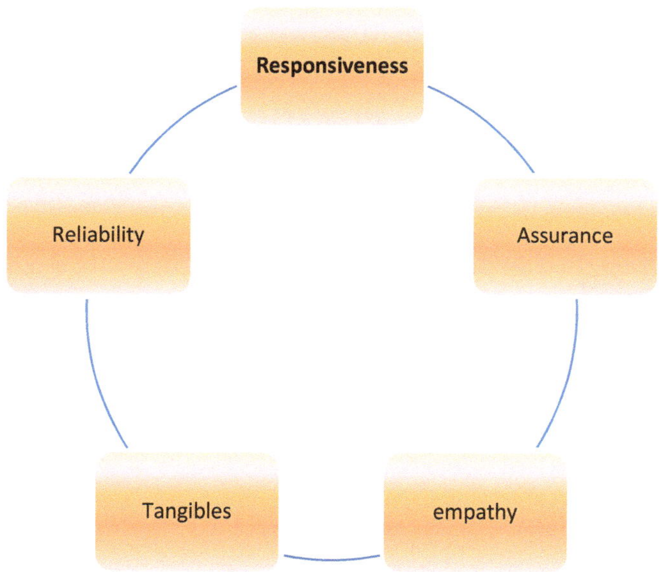

Figure No.3 Five service quality characteristics that please any customers. Discover desired qualities of service providers

2.3 BACKGROUNDAND REVIEWOF SERVICE QUALITY

Much of the initial work in developing a model to define and assess service quality has been conducted by Parasuraman Zeithaml and Berry (otherwise referred to as PZB). In conceptualizing the basic service quality model PZB identified ten key determinants of service quality as perceived by the company are the consumer reliability, responsiveness, competence, access, courtesy, communication, credibility, security, understanding/knowing the customer and tangibility [16].

It is noted that discrepancies existed between the firm's and the customer's perceptions of the service quality delivered. In investigating these discrepancies, PZB assert that service quality can be assessed by measuring the discrepancies or "gaps" between what the customer expects and what the consumer perceives he receives.

Arguing that the magnitude, and direction of this gap, directly affects the service quality that the consumer perceives, PZB note that customers would have perceptions of high service quality to the extent that their expectations are lower than the perceived service performance [16].

PZB assert that their framework can be used for identifying differences in the quality of goods and services by distinguishing between the properties of a good or service. They note that Nelson defined "search properties" as properties that can be determined before purchasing (such as credibility and tangibles), and "experience properties" as properties that can be determined only after purchase or consumption. Further, Darby and Karni defined "credence properties," (such as competence and security) as properties or characteristics that consumers often find extremely difficult to evaluate after their purchase. Therefore, PZB concluded that consumers typically rely on experience properties when evaluating service quality [16].

Based on their review of the literature, PZB developed the SERVQUAL scale. The scale was designed to uncover broad areas of good or bad service quality and can be used to show service quality trends over time, especially when used with other service quality techniques.

The SERVQUAL scale is based on a difference score between customer expectations of service and their perceptions after receiving the service. Initially PZB focused on the ten determinants of service quality.

3. MODEL OF SERVICE QUALITY GAPS

There are seven major gaps in the service quality concept, which are shown in Figure 1. The model is an extension of Parasuraman *et al.* (1985). According to the following explanation (ASI Quality Systems, 1992; Curry, 1999; Luk and Layton, 2002), the three important gaps, which are more associated with the external customers, are Gap1, Gap5 and Gap6; since they have a direct relationship with customers.

Gapl: Customers' expectations versus management perceptions: as a result of the lack of a marketing research orientation, inadequate upward communication and too many layers of management.

Gap2: Management perceptions versus service specifications: as a result of inadequate commitment to service quality, a perception of unfeasibility, inadequate task standardization and an absence of goal setting.

Gap3: Service specifications versus service delivery: as a result of role ambiguity and conflict, poor employee-job fit and poor technology-job fit, inappropriate supervisory control systems, lack of perceived control and lack of teamwork.

Gap4: Service delivery versus external communication: as a result of inadequate horizontal communications and propensity to over-promise.

Gap5: The discrepancy between customer expectations and their perceptions of the service delivered: as a result of the influences exerted from the customer side and the shortfalls (gaps) on the part of the service provider. In this case, customer expectations are influenced by the extent of personal needs, word of mouth recommendation and past service experiences.

Gap6: The discrepancy between customer expectations and employees' perceptions: as a result of the differences in the understanding of customer expectations by front-line service providers.

Gap7: The discrepancy between employee's perceptions and management perceptions: as a result of the differences in the understanding of customer expectations between managers and service providers.

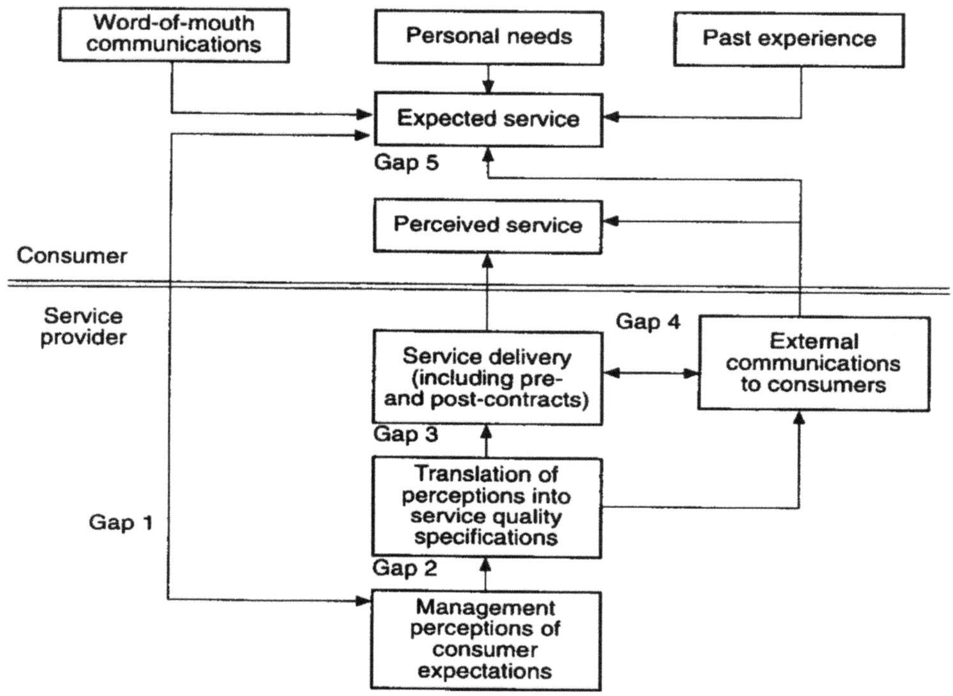

Figure No.4 Model of Service Quality Gaps

4. SERVQUAL METHODOLOGY

Clearly, from a Best Value perspective the measurement of service quality in the service sector should take into account customer expectations of service as well as perceptions of service. However, as Robinson (1999) concludes: "It is apparent that there is little consensus of opinion and much disagreement about how to measure service quality". One service quality measurement model that has been extensively applied is the SERVQUAL model developed by Parasuraman *et al* . (1985, 1986, 1988, 1991, 1993, 1994; Zeithaml *et al.* , 1990)[16-22]. SERVQUAL as the most often used approach for measuring service quality has been to compare customers' expectations before a service encounter and their perceptions of the actual service delivered (Gronroos, 1982; Lewis and Booms, 1983; Parasuraman *et al.* , 1985). The SERVQUAL instrument has been the predominant method used to measure consumers' perceptions of service quality. It has five generic dimensions or factors and are stated as follows (van Iwaarden *et al.* , 2003):[23]

(1) ***Tangibles***. Physical facilities, equipment and appearance of personnel.

(2) ***Reliability***. Ability to perform the promised service dependably and accurately.

(3) ***Responsiveness***. Willingness to help customers and provide prompt service.

(4) ***Assurance*** **(including competence, courtesy, credibility and security)**. Knowledge and courtesy of employees and their ability to inspire trust and confidence.

(5) ***Empathy*** **(including access, communication, understanding the customer).** Caring and individualized attention that the firm provides to its customers.

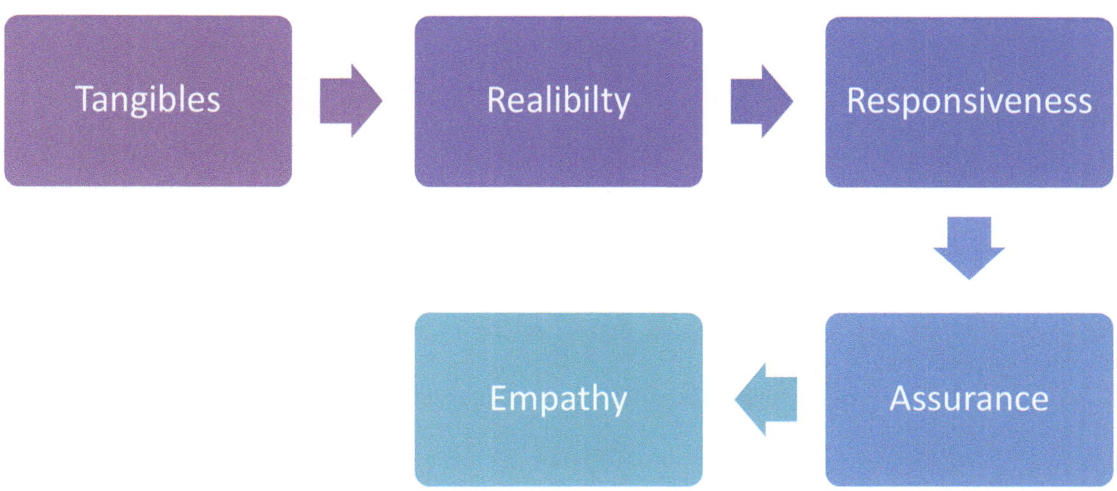

Figure No.5 five generic dimensions or factors

In the SERVQUAL instrument, 22 statements (Appendix I) measure the performance across these five dimensions, using a seven point likert scale measuring both customer expectations and perceptions (Gabbie and O'neill, 1996) [24]. It is important to note that without adequate information on both the quality of services expected and perceptions of services received then feedback from customer surveys can be highly misleading from both a policy and an operational perspective. In the following, the application of SERVQUAL approach is more specified with an example in a catering company.

5. DISCUSSION

The research on measuring service quality has focused primarily on how to meet or exceed the external customer's expectations, and has viewed service quality as a measure of how the delivered service level matches consumer's expectations. These perspectives can also be applied to the employees of a firm and in this case, other major gaps could be closed in the service quality gaps model.

6. CONCLUSION

In this paper, service quality and its model of gaps were reviewed. SERVQUAL attitude as an analytical approach for evaluating the dissimilarity between customers' expectations and perceptions of quality was also studied. While this research provides some perspectives to the field of service quality, it is supposed that there are a number of belongings that should be done to corroborate the demonstrated methodologies as well as to enhance the use of SERVQUAL in design and improvement of quality services in tractor industries.

Just as the SERVQUAL instrument is extensively used to weigh up external service quality, the instrument can also be customized to weigh up the quality of the internal service provided by departments and divisions within a company to employees in other departments and divisions. The results of the current study demonstrate that organizations can at least consider five dimensions of service quality to ascertain the level of services provided, and to determine which dimensions need improvement.

7. REFERENCES

[1]. R. D. Grisso, D. L. Morgan, G. J. Shropshire, S. K. Rockwell "What Information Helps a Farmer Purchase a Tractor?", 'APPLIED ENGINEERING in AGRICULTURE'.

[2]. Bedi Kanishka: Quality Management:Oxford University Press (2010).

[3]. http://en.wikipedia.org/wiki/Eight_dimensions_of_quality

[4]. Wisniewski M, Wisniewski H (2005). Measuring service quality in a hospital colposcopy clinic. Int. Journal Health Care Qual. 18(2/3): 217-229.

[5]. Lewis and Mitchell, 1990; Dotchin and Oakland, 1994a; Asubonteng et al., 1996; Wisniewski and Donnelly, 1996).

[6]. Parasuraman, A., Zeithaml, V.A. & Berry, L.L. (1985) A conceptual model of service quality and its implications for future research, Journal of Marketing, 49, pp. 41–50.

[7]. Woodside, A.G., L.L. Frey, and R.T. Daly (1989). 'Linking Service Quality, Customer Satisfaction, and Behavioral Intention'. Journal of Health Care Marketing. 9(4), p.5-17

[8]. Amitava Mitra , Quality Control and Improvement 2009.

[9]. Parasuraman, A., Zeithaml, V.A., & Berry, L.L, (1985), "A conceptual model of service quality and its implications for future research." Journal of Marketing, Vol. 49, fall.

[10] Parasuraman, A., Zeithamal, V.A., and Berry, L.L., (1988), "SERVQUAL: A Multiple- Item Scale for Measuring Consumer Perceptions of Service quality." Journal of retailing, volume 64, number 1, spring.

[11] Johnston R (1995). The determinants of service quality satisfiers and dissatisfies. Int. J. Serv. Ind. Manage., 6(5): 53-71.

[12] Grönroos, C. (2000), Service management and Marketing-A customer Relationship management approach.

[13] Zeithaml, V. A., Berry, L. L. & Parasuraman, A. (1985). "The behavioral consequences of service quality". Journal of Marketing, 60 (1), 31-46.

[14] Oppewal, H. & Vriens, M. (2000), "Measuring perceived service quality using integrated conjoint experiments", International Journal of Bank Marketing, Vol.18, No.4, PP. 154-69.

[15] Bahia, K. and Nantel, J. (2000), ``Areliableand valid measurement scale for the perceived service quality of banks'', International Journal of Bank Marketing, Vol.18 No.2, pp.84-91.

[16] Parasuraman, A., Zeithaml, V.A., &Berry, L.L, (1985), "A conceptual model of service quality and its implications for future research." Journal of Marketing, Vol. 49, fall.

[17] Parasuraman, A., Zeithaml, V.A. and Berry, L.L. (1985), "A conceptual model of service quality and its implication", *Journal of Marketing* , Vol. 49, Fall, pp. 41-50.

[18]Parasuraman, A., Zeithaml, V.A. and Berry, L.L. (1986), "SERVQUAL: a multiple-item scale for measuring customer perceptions of service quality", *Report No. 86-108*, Marketing Science Institute, Cambridge, MA.

[19]Parasuraman, A., Zeithaml, V.A. and Berry, L.L. (1988), "SERVQUAL: a multi-item scale for measuring consumer perceptions of the service quality", *Journal of Retailing* , Vol. 64, No. 1, pp. 12- 40.

[20]Parasuraman, A., Zeithaml, V.A. and Berry, L.L. (1991), "Refinement and reassessment of the SERVQUAL scalel", *Journal of Retailing* , Vol. 67, pp. 420-450.

[21]Parasuraman, A., Zeithaml, V.A. and Berry, L.L. (1993), "Research note: more on improving service quality measurement", *Journal of Retailing* , Vol. 69, No. 1, pp. 140-147.

[22]Parasuraman, A., Zeithaml, V.A. and Berry, L.L. (1994), "Reassessment of expectations as a comparison standard in measuring service quality: implications for future research", *Journal of Marketing* , Vol. 58, pp. 111-124.

[23] Van Iwaarden, J., van der Wiele, T., Ball, L., and Millen, R. (2003), "Applying SERVQUAL to web sites: An exploratory study", *International Journal of Quality & Reliability Management* , Vol.20, No.8, pp. 919-935..

[24] Gabbie, O. and O'Neill, M.A. (1996) SERVQUAL and the northern Ireland hotel sector: A comparative analysis – Part 1, *Managing Service Quality*, Vol.6, No.6, pp. 25-32.

5

THE ELEMENT AND MODIFICATION OF BEVEL GEAR

[1]Satish Kumar Patel, [2]Rajendra Kumar Patre, [3]Shankar Dayal Patel, [4]Prakash Kumar Sen., [5]Shailendra Kumar Bohidar

[1,2,3]Student,Bachelor of Engg. Mechanical Engineering
Kirodimal Institute of Technology, Raigarh C.G.
Raigarh, Chhattisgarh, INDIA
[4,5]Faculty,Mechanical Engg. Department
Kirodimal Institute of Technology, Raigarh C.G.
Raigarh, Chhattisgarh, INDIA

Abstract

Thermal deformation and other factor, the bevel gear pair will appear serious phenomenon of meshing interference and impact and uneven load in meshing process, so it is necessary to design suitable modification tooth to improve gear pairs transmission performance and enhance its bearing capacity. A methodology, which facilities to analyze the operating of bevel gears under consideration of production-caused deviations. The bevel gears are widely used in aerospace, marine and other field.

Keywords:-Bevel gear, gear teeth, gear drive.

I. Introduction

Bevel gears are widely employed in helicopter, automobiles, and engineering machinery for transmitting rotation and torque between intersected axis. The most important criteria for the quality meshing contact of bevel gear are low noise, and sufficient dimension and proper location of bearing contact. For the bevel gear with a conjugate gear tooth surface, when they are meshed under the condition of misalignments, these will be a relative motion and a large acceleration at the transfer point [1].

Because of its accurate transmission ratio, high transmission efficiency and big range of transmission power gear mechanism has become the most applied transmission mechanism [2]. Bevel gears are widely used in automobile differential and other parts of intersecting axis transmission. Its transmission quality will affect working performance of the part.

There are many design of gear which is manufactured depending upon their functionally, system requirement and operating condition. These includes spur gear, hypoid gear, spiral and straight bevel gear.[3]. Many gear failure occur due to design errors, manufacturing faults, maintenance, inspection, inevitable respective stresses resulting in surface fatigue wear and deterioration of lubrication properties. In this study the cause of bevel gears in an engine train of an aircraft has been investigated to avoid future [4].

II. The Modification of Bevel Gear

In order to avoid "edge effect", alleviate the meshing impact and reduce the influence degree of the transmission performance of bevel. Gear due to alternating load and assembly error. There are two kinds of modification method symmetric crowned modification and isometric modification to design tooth finally by comparing the comprehensive effect of these two modification methods.

II (I). Symmetric Crowned Modification

Symmetrical crowned modification was adopted only for planetary gear and the drum point was in the middle of tooth wide. The modification amount was determined by certain condition. The recommended symmetrical crowned modification gears formulas are follows [5].

$$Rc = \frac{b^2}{8\Delta}$$

Where, Rc are arc radius,
b is tooth width &
is amount of symmetric crowned modification.

II (II). Isometric Modification

Isometric modification was adopted only for planetary gear that is along with the planetary gear original tooth surface normal direction equidistant stretched out new tooth surface which paralleled to original tooth surfaces the part of isometric modification maintenance the characteristic of involutes surface.[6,7].

III. Methodology for Requirement- Driven Tolerance Specification of Bevel Gear

Starting with the current product development process of bevel gears, a methodology was developed to analyze the complex interrelationship mentioned in the first chapter.

In addition to functional requirements, a user- friendly environment, the effort of time, the independence from existing software solution and the possibility for automation are important criteria for the implementation of a new methodology reveals a modular structure which can be seen on the process diagram below (fig.1)

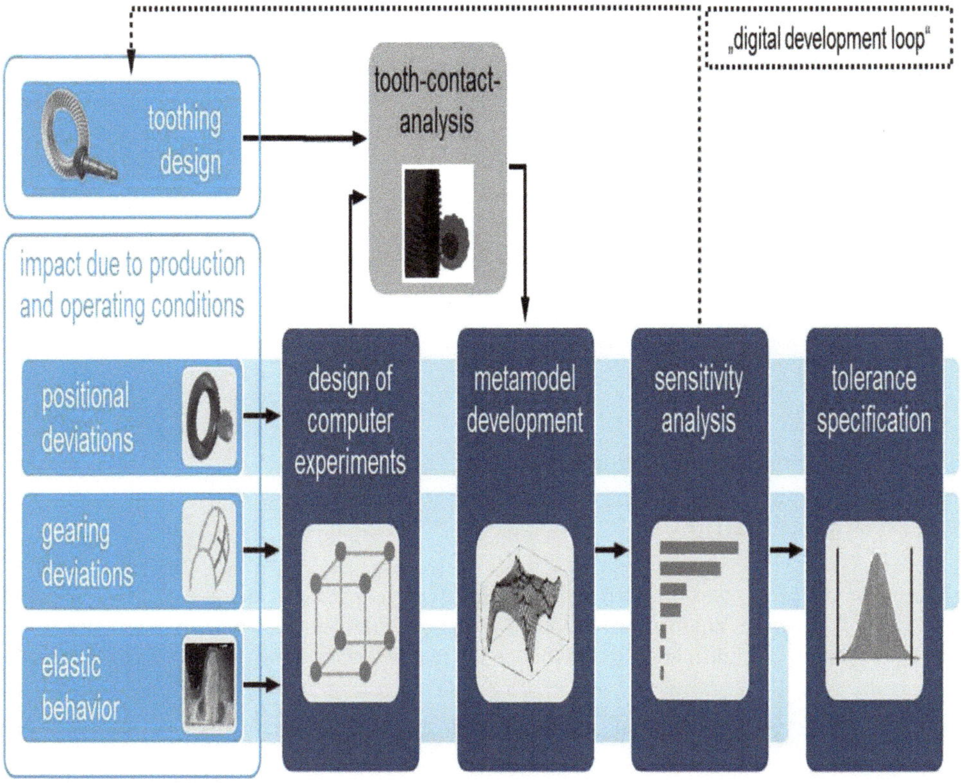

Fig.1. Methodology for requirement-driven tolerance specification of bevel gear.[8]

Based on a given tooth, tooth contact for specific combination of deviation is analyzed. Input parameter can be categorized into positional deviation of the pinion and the ring gear, topographical deviation of the tooth and elastic behavior of the parts. Design of computer experiments is used to maximize the amount of information gained from a minimize number of experimental runs.[8].

IV. VISUAL OBSERVATION

The damaged Drive gear teeth are numbered accordingly and schematic view is presented in (fig.2). It was observed that drive gear was severely damage by total loss of approximately 9 teeth and the rest 11 were badly deformed with cracked finned edge at the tips. The surface was contaminated with debris and lubricating oil. The cracks and pitting at the deformed teeth surface occurred due to the cold working of case under higher applied loads. The drive gear teeth (2,3,4,5,6,7) were badly worn and severely case was sheared off during sliding of meshing gear in (fig3).[4].

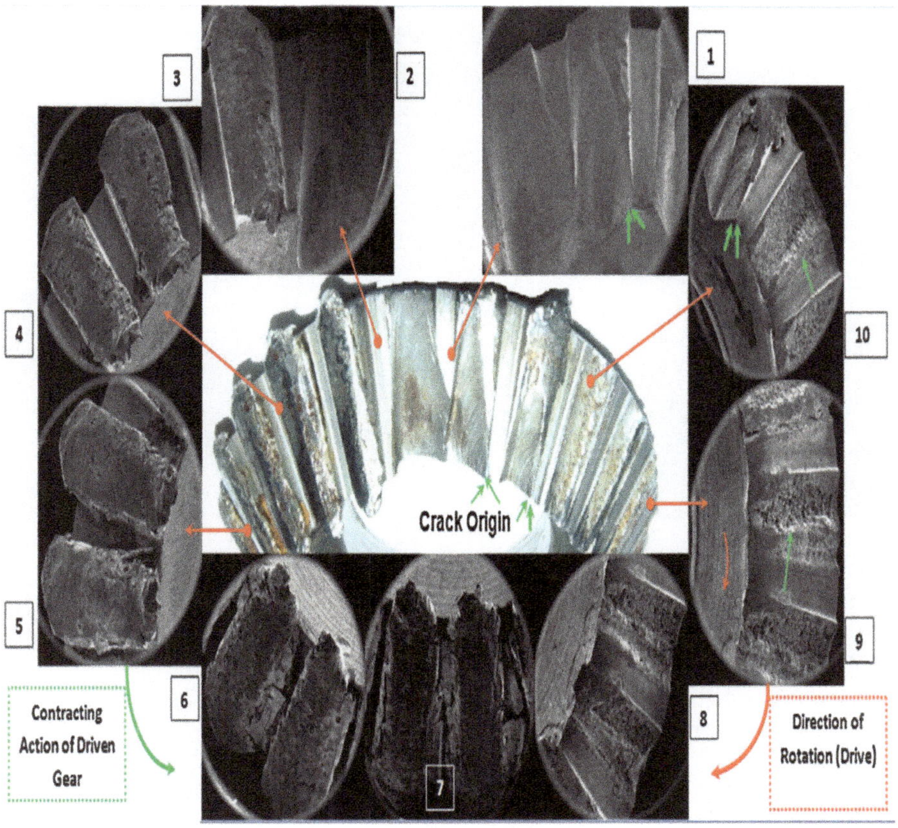

Fig. 2 Schematic view of damaged drive gear (pairs of teeth are numbered from 1–10).[4]

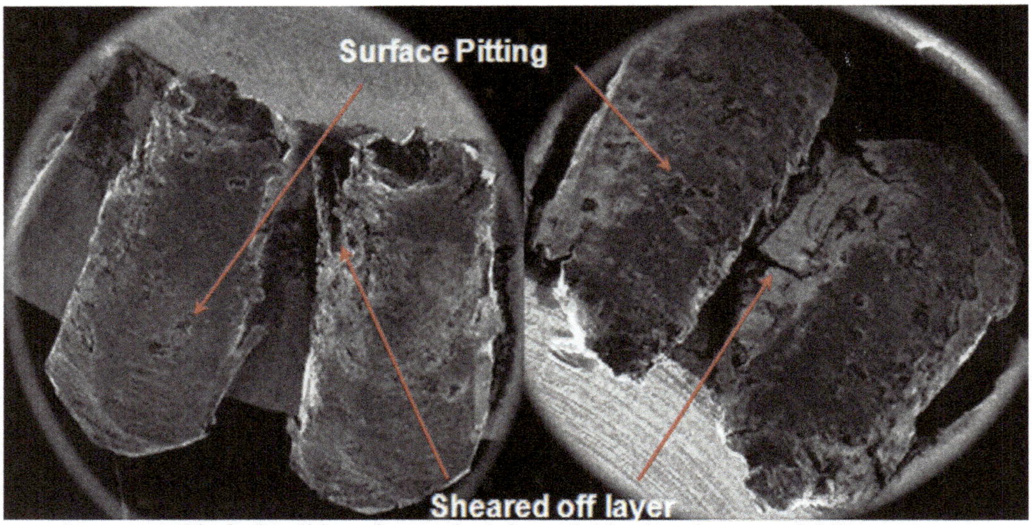

Fig 3. Condition of severely deformed drive bevel gear teeth.[4]

V. Chemical Composition of Bevel of Gear

Chemical composition of Drive and Driven Bevel Gear is given in Table 1.The composition of gear confirmed to AISI 8617 grade. No significant variation n composition was observed in both gears samples,

V (I).SURFACE ANALYSIS (EDS)

The surface of drive gear was analyzed by (energy dispersive spectroscopy) EDS for contamination and any compositional variation; the result in (fig,4). The surface was enriched with carbon validating the surface hardening through.

Table.1Chemical composition of Driven gear.

Elements (wt %)	Drive gear	Driven gear
Carbon (C)	0.18	0.19
Manganese (Mn)	0.85	0.82
Silicon (Si)	0.24	0.24
Chromium (Cr)	0.58	0.56
Molybdenum (Mo)	0.21	0.20
Nickel (Ni)	0.30	0.4
Sulfur (S)	0.02	0.052
Iron (Fe)	Balance	Balance

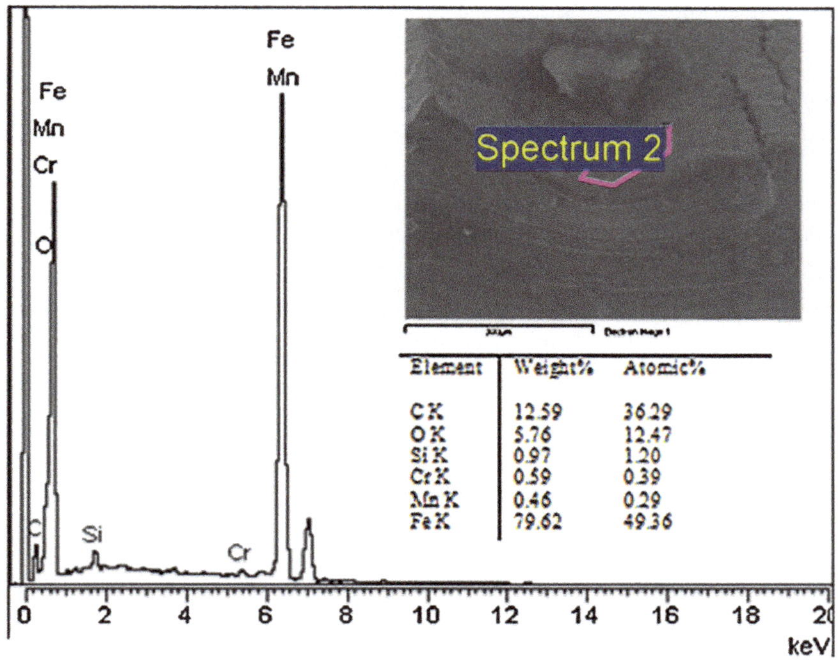

Fig. 4 EDS analysis of drive gear.

Carburizing process. The higher concentration of oxygen (O) and silicon (Si) at the surface of gear revealed the oxidation and contamination which may produce localized boundary lubrication condition (BLC) between the meshing gears and hence may cause wear and pitting. The increase in temperatures due to BLC caused the degradation of lubrication properties further and surface oxidation took place as depicted qualitative in EDS spectrum from high concentration of carbon and oxygen. This situation was aroused just after 81.2 flight hours since last total oil charge. The very thin layer of lubrication also promotes non uniform distribution of stress which may cause fracture of gear teeth [4].

VI. Conclusion

Analyzing the result of dynamic contact simulation and comparing the comprehensive of symmetric crowned modification with isometric modification. The conclusion can be made that these two modification method can all improve the meshing impact and avoid " end contact " in meshing process of gear pair. But the improvement of comprehensive properties of isometric modification is discussed above is more effective than symmetric crowned modification for bevel gear [9].

REFERENCES

[1].So Jinzhan ,Frang Zongde, Cai Xianwei ohines Journals of Aeronatics (2013), 26 (5);1310-1316.

[2].Sun H, Chen Zm: Theory of machines and mechanisms. Beijing : Higher Erucation Press; 19995.

[3].Ramachandran V, Raghuram AC, Krishnan RV, Bhaumerhodologies and case histories. In: materials park, ohio: Asm international; 2005.

[4]. Nauman A. Siddiqui, KM. oein, M. Zuban bhan, R. Ahmad case studies in engineering failure Analysis 1 (2013) 24-31.

[5]. AIDA. Design and Manufacturing of gear. Beijing; Agriculture machinery press of china; 1983.

[6]. Cui HY, Yang D, Tian XJ, etal. Research on Axial modification of Bevel Gear machine Design and research 20[1]1; 27 (2) ; 33-36.

[7]. Tian XJ. Research on tooth - modification Technology of fine- forming straight - tooth Bevel Gear. School of materials science , Engineering, Shandong University; 2009.

[8]. Monfgomery DC. Design and Analysis of Experiments. 7th ed. Hobolcen, NJ: wiley, 2009,-inteenational student version.

[9]. Fangyou Zhang' , Xije Tian, Huanyong Cui. IERI procedia 3 (2012) 52-59.

A Review on "Finite Element Analysis of Chassis using ANSYS"

SUPREET MAHADEOKAR[1], ANSHUL CHOUDHARY[2]

[1]*Research Scholar, Department of Mechanical Engineering, SRIT, Jabalpur (M.P.) India*

[2]*Assistant Professor, Department of Mechanical Engineering, SRIT, Jabalpur (M.P.) India*

Abstract

At present review paper an effort is made to study the previous investigations that have been made on the different analysis techniques of automobile frames. That analysis may be fatigue analysis, static analysis or dynamic analysis. A number of analytical and experimental techniques are available for the analysis of the automobile frames. Determination of the different analysis around different condition through the various methods lie FEA, in an automobile frames has been reported in literature.

Keywords: FEA; ANSYS, Chasis.

1. Introduction

In reading tests of and articles about new cars in newspapers and magazines and on The Auto Channel, we have seen much attention paid to increased chassis rigidity in new vehicles. We on the writing-about-it side certainly hear much on the subject, at press conferences and new car introductions and in the informational material given to us by the auto manufacturers. "The all-new 1998 Bloxfire GT is 75% more rigid than the 1997 model" is a typical sort of statement. "Whatever does that mean and is it important?"

It certainly is important. The chassis of a car (or truck, or any other sort of wheeled vehicle) is the most important structural part. The more rigid it is, the better chance that all of the wheels will point in the desired direction. This is a Good thing for control, safety, and comfort.

Consider control, primarily steering response and cornering ability. If the vehicle's frame flexes too much, the wheels, and therefore the vehicle itself, will move in directions other than the one which the driver intended. Not a Good Thing.

This movement doesn't have to be large to be noticeable. The need for constant steering corrections while driving may be due to frame flex (among many other possible causes.) This can be tiring at best (a definite safety factor), and can make a car's cornering abilities less than optimum - another very negative safety factor. In the "good olde days" when automobile frames were less rigid than today, sports cars often had very stiff suspensions. This reduced unwanted wheel movement. It also reduced comfort, and, seemingly paradoxically, sometimes reduced handling abilities as well. No paradox, really. The tires must be in contact with the ground in order to transmit acceleration, braking, or cornering forces. A too-stiff suspension will have wheels rebounding into the air, where the tires do nothing.

Another noticeable effect of flex is noise. As different parts of a vehicle move with respect to each other, noise results. With time, things loosen and get even more annoying noisy. Squeaks and rattles are the most noticeable effects, but constant flex or vibration can break things, too. Not good, for cosmetic, comfort, or possibly (depending on what breaks) safety reasons.

So the modern, more rigid car is quieter, more comfortable, handles better, and is safer than its flaxy forebear. The automobile as we know it is barely over 100 years old, and depends on quite a few other technologies as well. There has been a certain amount of mechanical natural selection over the past century, with plenty of evolutionary dead ends and side tracks along the way. Many modern materials, construction techniques, and auxiliary technologies used in automobile manufacture today are considerably newer than that century, too. Karl Benz and Henry Ford didn't have the fruits of modern metallurgy, heavy-duty machine tools, space-age composite materials, or computers. They barely had electricity. Industry pioneers adapted horse-drawn wagon and bicycle technology to the automobile, but the automobile is a different sort of machine than either of those. It took time to learn, and they're still learning.

Chassis Frame: Chassis is a French term and was initially usedto denote the frame parts or Basic Structure of the vehicle. It is the back bone of the vehicle. A vehicle with-out body is called Chassis. The components of the vehicle like Power plant, Transmission System, Axles, Wheels and Tyres, Suspension, Controlling Systems like Braking, Steering etc., and also electrical system parts are mounted on the Chassis frame. It is the main mounting for all

the components including the body. So it is also called as Carrying Unit.

The following main components of the Chassis are

1. Frame: it is made up of long two members called side membersriveted together with the help of number of cross members.

2. Engine or Power plant: It provides the source of power

3. Clutch: It connects and disconnects the power from the engine flywheel to the transmission system.

4. Gear Box

5. U Joint

6. Propeller Shaft

7. Differential

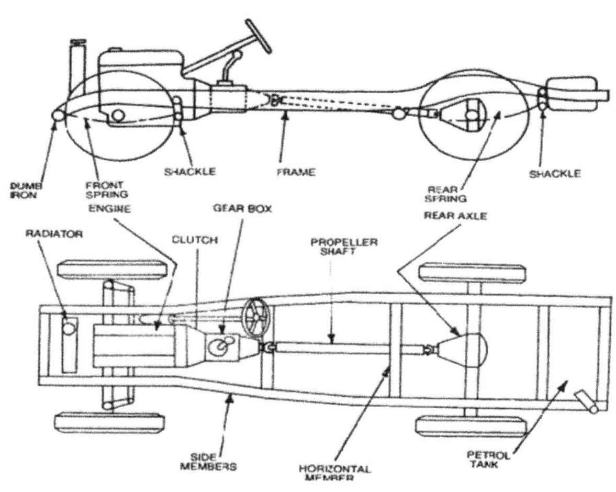

Fig 1.1 a general Layout of Chassis and its main Components

Functions of the chassis frame:

1. To carry load of the passengers or goods carried in the body

2. To support the load of the body, engine, gear box etc.,

3. To withstand the forces caused due to the sudden braking or acceleration

4. To withstand the stresses caused due to the bad road condition.

5. To withstand centrifugal force while cornering

Types of chassis frames:

There are three types of frames

1. Conventional frame

2. Integral frame

3. Semi-integral frame

1. Conventional frame: It has two long side members and 5 to 6 cross members joined together with the help of rivets and bolts. The frame sections are used generally.

a. Channel Section - Good resistance to bending

b. Tabular Section - Good resistance to Torsion

c. Box Section - Good resistance to both bending and Torsion

2. Integral Frame:

This frame is used now a day in most of the cars. There isno frame and all the assembly units are attached to the body. All the functions of the frame carried out by the body itself. Due to elimination of long frame it is cheaper and due to less weight most economical also. Only disadvantage is repairing is difficult.

3. Semi - Integral Frame:

In somevehicles half frame is fixed in the front endon which engine gear box and front suspension is mounted. It has the advantage when the vehicle is met with accident the front frame can betaken easily to replace the damaged chassis frame. This type of frame is used in FIAT cars and some of the European and American cars.

Various loads acting on the frame:

Various loads acting on the frame are

1. Short duration Load - While crossing a broken patch.

2. Momentary duration Load - While taking a curve.

3. Impact Loads - Due to the collision of the vehicle.

4. Inertia Load - While applying brakes.

5. Static Loads - Loads due to chassis parts.

6. Over Loads - Beyond Design capacity.

Fatigue, static & Dynamic Analysis

Fatigue damage concepts:

Stress-Life Diagram (S-N Diagram): The basis of the Stress-Life method is the Wohler S-N diagram, shown schematically for two materials in Figure 2.The S-N diagram plots nominal stress amplitude S versus cycles toFailure N. There are numerous testing procedures to generate the required data for a proper S-N diagram. S-N testdata are usually displayed on a log-log plot, with the actual S-N line representing the mean of the data from several tests.

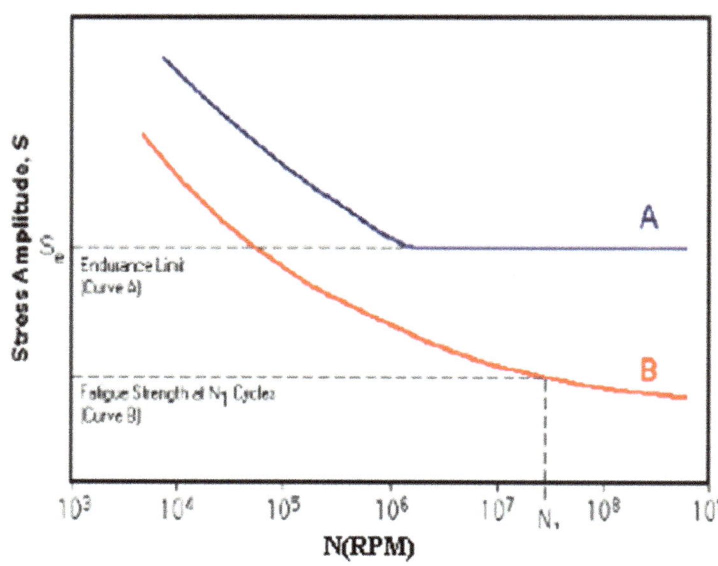

Fig 2 S-N Curve

Review on previous investigations

CicekKaraoglu and ***N.SefaKuralay*** did the finite element analysis of a truck chassis. The analysis showed thatincreasing the side member thickness can reduce stresses on the joint areas, but it is important to realize that the overall weight of the chassis frame increases. Using local plates only in the joint area can also increase side member thickness. Therefore, excessive weight of the chassis frame is prevented.

In November 2008 ***MohamadTarmizi Bin Arbain*** Mechanical Engineering department Universiti Malaysia Pahang he use 3D model for finite element analysis issues regarding the experimental analysis of car chassis is addressed. The modeling approach is investigated extensively using both of computational and compared it to experimental modal analysis.

A comparison of the modal parameters from both experiment and simulation shows the validity of the proposed approach. Then perform the computational stress analysis with linear material type analysis to find the stress concentration point in thecar chassis.

Karaoglu and ***Kuralay*** investigated stress analysis of a truck chassis with riveted joints using Fem. Numerical results showed that stresses on the side member can be reduced by increasing the side member thickness locally. ***Fermer et al***investigated the fatigue life of Volvo S80 Bi-Fuel using MSC/Fatigue Conle and Chu did research about fatigue analysis andthe local stress –strain approach in complex vehicular structures. Structural optimization of automotive components applied to durability problems has been investigated by Ferreira et al

FilhoEt. al. have investigated and optimized a chassis design for an off road vehicle with the appropriate dynamic and structural behavior.In1837, Wilhelm Albert publishes the first article on fatigue.

***Manpreetsinghbajwa* et.al.(2013)** reported onthe effect of static load and give an analysis (excluding damping and inertia effects) of the chassis of TATA super ace using ANSYSworkbench and itsverification using solid mechanics. The chassis of TATA super ace is of ladder frame type which has two side members or longitudinal members of C- cross section and five transverse members called as cross members of box cross section. The chassishas been modelled in CATIA V5R18 using the most of the actual dimensions. FEM analysis was done using ANSYS 14 workbench. Dimensions used for analysis

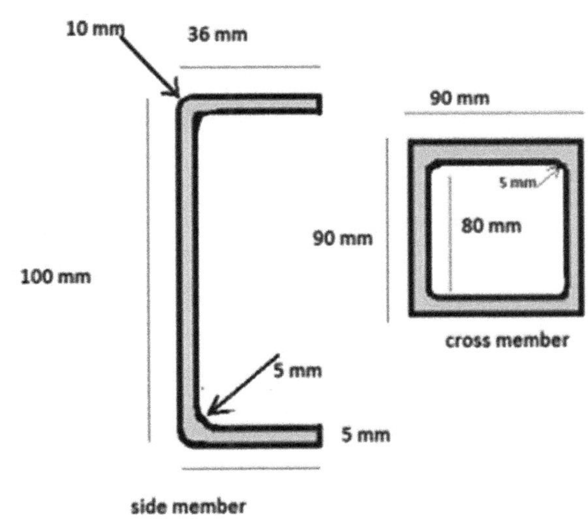

Fig 3 Dimensions of chassis using by *Manpreetsinghbajwa* et.al.(2013)

Finally the design, static analysis and its verificationusing solid mechanics has been successfully accomplished. The work not only provides an insight into the design and analysis of the chassis but alsopinpoints the critically stressed points where the design can be modified for improving the chassis.

Swami K.I. et.al.(2014) related his work to perform towards the static structural analysis of the truck chassis. Structural systems like the chassis can be easily analyzed using the finite element techniques. Thus proper finite element model of the chassis has been developed. The chassis is modeled in ANSYS. Analysis is done using thesame software.

SanikaOturkaret. al.(2013)to give an introduction to this staticstress analysis method using ANSYS APDL 12.0, covering topics such as impact force determination, loading points, convergence of nodes, and the mesh size dependence of generated stress.The analysis highlights the areas of high stress concentration, which need a change in design. This preliminary analysis consumes little time and approves of a reasonable design, which can form the basis for a detailed

Modeled.

Andrew Salzano et.al.(2009)focused on the design, analysis, and fabrication of a steel tube space frame chassis for use in Formula SAE. A guideline for the construction of this frame will be based upon the Wolfpack Motorsports 2009 car (WMF09).

Dr.R.Rajappan et.al.(2013)focused onstatic and dynamic load characteristics of chassis which are analysed using FE models from this work. It is found that some location of high stress area, analyzing vibration, natural frequency and mode shape by using finite element method. Modal updating of truck chassis model will be done by adjusting the selective properties such as mass density and Poisson's ratio. Predicted natural frequency and mode shape has been validated against previously published result. Finally, the modifications of the updated FE truck chassis model has been proposed to reduce the vibration, improve the strength, and optimize the weight of the truck chassis.

4. Conclusion

In this paper an effort is made to review the previousinvestigations that have been made on the analysis of variousautomobile frames structure. An attempt has been made in the present article to give an overview of various techniques developed for the analysis of automobile frames and results of that analysis due to which further study on the chassis will become easy. An information of assessment of a suspension arm, vehiclesuspension components, analysis of truck chassis, for the fatigue analysis of a truck chassis with riveted joints are considered.

References

[1] Stress Analysis Of Heavy Duty Truck Chassis As A Preliminary Data Using Fem, RoslanAbdRahman, MohdNasirTamin, Journal Mekanikal, December 2008, No.26, 76 – 85.

[2] Statics and Dynamics Structural Analysis Of A 4.5 Ton Truck Chassis, Teo Han Fui, RoslanAbd. Rahman*, Journal Mekanikal, December 2007, No. 24, 56 – 67

[3] Stress analysis of a truck chassis with riveted joints, CicekKaraoglu, N. SefaKuralay, Journal of Finite Elements in Analysis and Design 38 (2002), Elsevier Science, page no- 1115–1130.

[4] Karaoglu, C. and Kuralay, N.S., 2000. Stress Analysis of a Truck Chassis with Riveted Joints, Elsevier Science Publishers B.V.Amsterdam, the Netherlands, Vol. 38, 1115-1130.

[5] Conle, F.A. and Chu, C.C., 1997. Fatigue Analysis and the Local Stress-strain Approach in Complex Vehicular Structures, International Journal of Fatigue.

7

EXPERIMENTAL STUDY ON PERFORMANCE AND EMISSION CHARACTERISTICS OF DIESEL ENGINE USING RICE BRAN OIL AS FUEL

Anandkumar S Malipatil[1], Manjunath Bandi [2]

[1] *VTU* , *jovemp.aks@gmail.com*

[2] *VTU,* manjunath.bandi05@gmail.com

CorrespondingAuthor:VTU,Gulbarga,jovemp.aks@gmail.com

Abstract

Diesel engine is a prime mover which is widely used where the electric power is not available and for transportation. Due to depletion of fossil fuels, it is necessary to look after the alternate fuels. The use of Biofuels can partially fulfill the problem of fuel crises and environmental issues. Biofuels are classified into bio-ethanol and biodiesel. Biodiesel is a monoalkyl ester of long chain fatty acids synthesized from vegetable oils (Edible and Non-edible oils).Bio-fuel is a clean burning fuel made from natural renewable energy resource; it operates on IC engine similar to the petroleum diesel. The rising cost of diesel and the danger caused to the environment has led to an intensive and desperate search for alternative fuels. Among them, the rice bran oil has proven to be a promising substitute to diesel. In this experimental study, A 4-stroke, single cylinder, constant speed, direct injection diesel engine was operated on rice bran oil-biodiesel of different blends. Four different blends of 20, 40, 60 and 80 % by volume were used for this study. Various engine performance, combustion and emission parameters such as Brake Thermal Efficiency, Brake Specific Fuel Consumption, Heat Release Rate, Peak Pressure, Exhaust Gas Temperature, etc. were recorded from the acquired data. The recorded parameters were studied for varying loads and their corresponding graphs have been plotted for comparison purposes. Petroleum Diesel has been used as the reference. From the properties and engine test results it has been established that rice bran oil biodiesel is a better replacement for diesel without any engine modification.

Key words: Rice bran oil, biodiesel, diesel, diesel engine

1. Introduction

India is one of the fastest growing economies in the world. The Development Objectives focus on economic growth, equity and human wellbeing. Energy is a critical input for socio-economic development. The energy strategy of a country aims at efficiency and security and to provide access which being environment friendly and achievement of an optimum mix of primary resources for energy generation. Fossil fuels will continue to play a dominant role in the energy scenario in our country in the next few decades. However, conventional or fossil fuel resources are limited, non-renewable, polluting and, therefore, need to be used prudently. On the other hand, renewable energy resources are indigenous, non-polluting and virtually inexhaustible. India is endowed with abundant renewable energy resources. Therefore, their use should be encouraged in every possible way. The crude oil price has been fluctuating in the world market and has increased significantly in the recent past, reaching a level of more than \$ 140 per barrel. Such unforeseen escalation of crude oil prices is severely straining various economies the world over, particularly those of the developing countries. Petro-based oil meets about 95% of the requirement for transportation fuels, and the demand has been steadily rising. Provisional estimates have indicated crude oil consumption in 2007-08 at about 156 million tons. The domestic crude oil is able to meet only about 23% of the demand, while the rest is met from imported crude. India's energy security would remain vulnerable until alternative fuels to substitute/supplement petro-based fuels are developed based on indigenously produced renewable feedstocks. In bio fuels, the country has a ray of hope in providing energy security. Bio fuels are environment friendly fuels and their utilization would address global concerns about containment of carbon emissions. The transportation sector has been identified as a major polluting sector. Use of bio fuels has become compelling in view of the tightening automotive vehicle emission standards to curb air pollution. Bio fuels are derived from renewable bio-mass resources and, therefore, provide a strategic advantage to promote sustainable development and to supplement conventional energy sources in meeting the rapidly increasing requirements for transportation fuels associated with high economic growth, as well as in meeting the energy needs of India's vast rural population. Bio fuels can increasingly satisfy these energy needs in an environmentally benign and cost- effective manner while reducing dependence on import of fossil fuels and thereby providing a higher degree of National Energy Security. Liquid fuel meets most of the energy requirement of the transport sector with a continuous rise in the demand over the past decades across the country. The growth of the demand for the liquid fuel is increasing almost 30% over every 4-5 years. The crude oil supply from the domestic sources, hardly reach 30% of the demand, while the rest met out of the imports. In this project Bio-fuels provide a ray of hope to offset a reasonable margin of the imports serving as a good substitute for the fossil fuels. The present bio-fuels that are bio-diesel, methanol and ethanol which are gaining immense importance and awareness across the globe. This is particularly because of the wider acceptance due to its environment friendly, providing energy security at grass root level providing employment to rural community and improved income generation.

In 1984 Pryde ***et al.*** showed that the methanolysis of soybean oil, in the presence of 1 mol% of H2SO4, with an alcohol/oil molar ratio of 30:1 at 65°C, takes 50 h to reach complete conversion of the vegetable oil (> 99%), while the butanolysis (at 117 °C) and ethanolysis (at 78 °C), using the same quantities of catalyst and alcohol, take 3 and 18 h, respectively. **Freedman** ***et al in 1986***. found that a 30:1 ratio of methanol to soyabean oil with 1% sulphuric acid gave higher conversion after 44 h of heating at 60°C. they emphasized that if vegetable oils have more than 1% free fatty acids, the acid catalysis become more effective than alkali catalyst. **Sukumar** ***et al****.in2005* conducted an experiment on a direct injection 4 stroke diesel engine using methyl esters of MAHUA OIL and found that, engine performance with Mahua oil ester does not differ greatly from that of diesel fuel. Vegetable oil was used as an alternative fuel for diesel engine. **Lang** ***et al***.in**2006** have synthesized canola methyl ester (CME), rapeseed methyl ester (RME), linseed methyl ester (LME) and sunflower methyl ester (SME) in a batch type reactor using both potassium hydroxide and sodium hydroxide as catalysts and maintaining molar ratio of 6:1. It has been reported that in a commonly used single stage process, a period of 1 hr is required to achieve 98% conversion of rapeseed oil to rapeseed methyl ester **Balusamy** ***et al****.in 2007* have investigated methyl ester of thevetiaperuviana seed oil (TPSO) and blended with diesel fuel, has been tested in naturally aspirated single cylinder diesel engine a trated speed of 1500 rpm. **Kandasamy** ***et al****.in 2009* have discussed the performance characteristics of a single cylinder diesel engine using rice bran and pungam oil blended with diesel fuel.From the above review of literature the following important conclusions are made. The past work reveals that they used esters of edible and non-edible oil like soybean, honge, jatropha, yellow grease, and mahua oil in place of diesel in a single cylinder and heavy duty diesel engines. So far a very few kinds of vegetable oil have been tested on CI engines. India has the one of the great potential for production of bio-fuels from rice bran oil. Hence there is a need to identify new kinds of biodiesel such as rice bran oil biodiesel and examining their suitability as an alternate fuel in single cylinder 4-stroke diesel engines.

2. Methodlogy

Extraction Of Biodiesel From Rice Bran Oil By Transesterification Process

The transesterification is an equilibrium reaction and the transformation occurs essentially by mixing the reactants. However, the presence of a catalyst (typically a strong acid or base) accelerates considerably the adjustment of the equilibrium. In order to achieve a high yield of the ester, the alcohol has to be used in excess.

Mild Acid Catalyzed Transesterification

This is a type of reaction that takes place in the presence of methanol (30%) and orthophosphoric (0.6%) acid at 60°c with constant stirring, helps in the separation of impurities which were dissolved in the methanol as an upper layer and oil in the lower layer.

Strong Acid Catalyzed Transesterification

The first stage product is reacted with the sulphuric (0.6%) acid and methanol (20%) for 2 hours at 60°c with constant stirring. The reaction product is allowed to settle. The FFA and the other impurities were removed in this stage as an upper layer and oil in the lower layer. The product of earlier stages i.e. pure triglycerides is made to react with methanol (30%) and catalyst, KOH (3gms) for 2 hours at 60°C with constant stirring rate. The reacted product of this third stage is made to settle down under gravity. The lower contains which contain glycerol and other impurities are removed and further excess of alcohol and other impurities present are removed by water wash process after the pH neutralization. The water wash product then heated above 100°C in order to remove the moisture content.

3. Governing Equations

1. **Brake Power**

$$BP = \frac{2\pi NT}{60\times 1000}$$

2. **Specific Fuel Consumption**

$$BSFC = \frac{X1\times 3600}{t\times 1000\times BP}$$

3. **Specific Energy Consumption**

$$BSEC = \frac{X2\times 3600}{t\times 1000\times BP}$$

4. **Brake Thermal Efficiency (%)**

$$\eta_{bth} = \frac{BP\times t\times 1000}{X2}\times 100$$

Note:

1. $X1 = (\text{Volume} \times \text{Specific gravity})^{BD} + (\text{Volume} \times \text{Specific gravity})^{D}$

2. $X2 = (\text{Volume} \times \text{Specific gravity} \times CV)^{BD} + (\text{Volume} \times \text{Specific gravity } CV)^{D}$

Blending Of Fuel

The produced bio diesel is blended with the regular Diesel in different percentages. Below are the notations for different fuel samples which are blended.

Sl. No.	Bio diesel percentage	Diesel percentage	Notation
1	20%	80%	B20
2	40%	60%	B40
3	60%	40%	B60
4	80%	20%	B 80
5	100%	0%	B100

The blending process was carried out with the help of a measuring jar and beaker. The appropriate percentages of diesel and biodiesel were added to the beaker and then transferred to bottle. The bottles were shaking well and were allowed to stay upside down to ensure proper mixing of fuels. The bottles were stored in dry place and kept still for the next 24 hour. Blends were checked for every 6 hrs time intervals for any layer formation. All the blends were stable and passed the 24 hrs stability test and were ready to be used on engine.

4. Properties of Rice Bran Biodiesel

The blended fuel samples and biodiesel were tested for different chemical and physical properties. The results have been furnished below

CALORIFIC VALUE (KJ/Kg)	40164
DENSITY (Kg/m^3)	909
KINEMATIC VISCOSITY CST at 40°C	30
FLASH POINT ^{o}C	126
FIRE POINT ^{o}C	152

5. Experimental Setup for Engine Performance Test

A four stroke, single cylinder water cooled diesel engine is employed for the present study. The detail specification of the engine used are given in table. Five gas analyzer was used to measure the concentration of gaseous emissions such as Oxides of nitrogen, unburned hydrocarbon, carbon monoxide, carbon dioxide and oxygen level. The performance and emission tests are carried out on the C.I. engine using various blends of diesel-biodiesel blends as fuels. The tests are conducted at the constant speed of 1500rpm at various torque.

SLNO	PARAMETERS	SPECIFICATION
1	Type	TV 1 (kirloskar made)
2	Nozzle opening pressure	200 bar
3	Governor type	Mechanical centrifugal type
4	Number of cylinders	Single cylinder
5	Number of strokes	Four stroke
6	Fuel	Diesel
7	Compression ratio	16.5:1
8	Cylinder diameter(Bore)	80mm
9	Stroke length	110mm
	Electrical dynamometer	
10	Type	Foot mounted, continuous rating
11	Alternator rating	3KVA
12	Speed	2800-3000RPM
13	Voltage	220 V AC

6. Experimentation Methodology

First the experimentation is performed with diesel (for getting the base line data of the engine) and then blends of different percent volumes of Biodiesel B20, B40, B60, B80 were carried out. The performance of the engine is evaluated in terms of brake thermal efficiency, brake specific energy consumption, exhaust gas temperature, and emission of the engine is analyzed (HC, CO, CO2, O2 and NOx).

7. Results and Discussion

The studies were, conducted on blends of different present volumes of Biodiesel B20, B40, B60 and B 80 were carried out.

Engine Performance And Exhaust Emission Analysis

Performance Characteristics

The performance characteristics of the engine are the very important criterion for selection and suitability of alternate fuels. This study evaluates BSFC, BTE and EGT of diesel blends.

Brake Thermal Efficiency

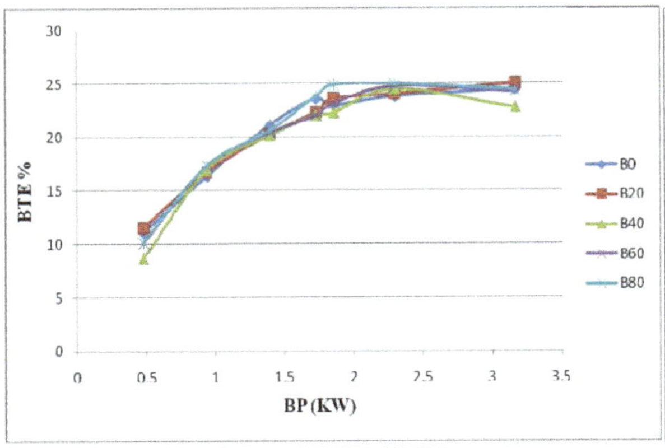

Fig.7.1Variation of BTE with BP

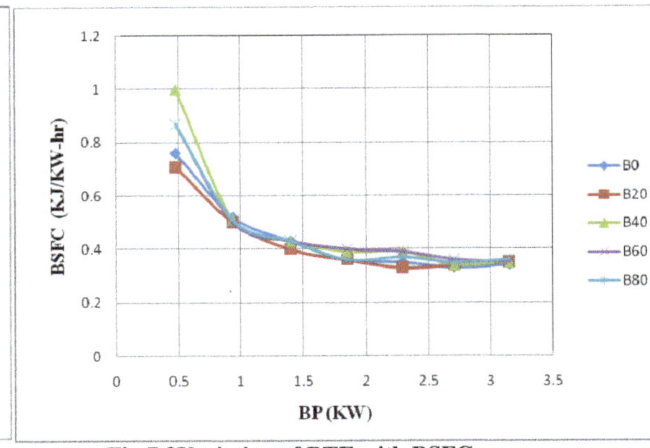

Fig.7.2Variation of BTE with BSFC

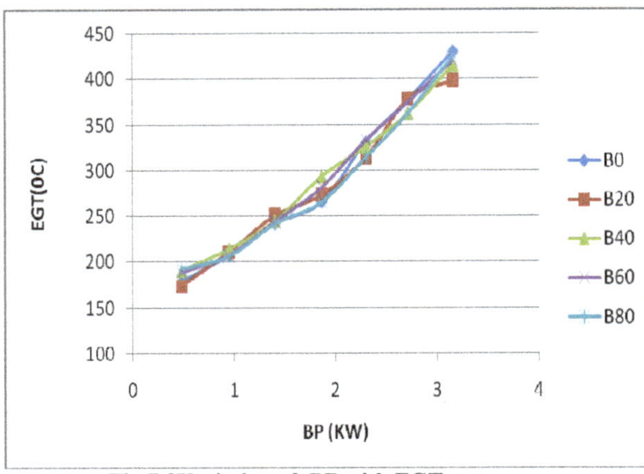

Fig.7.3Variation of BP with EGT

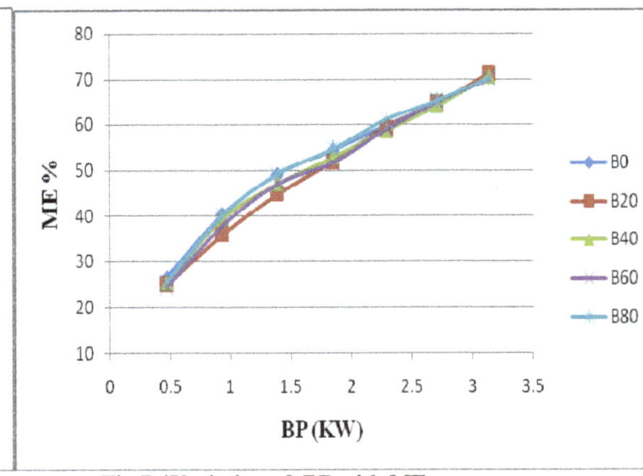

Fig.7.4Variation of BP with ME

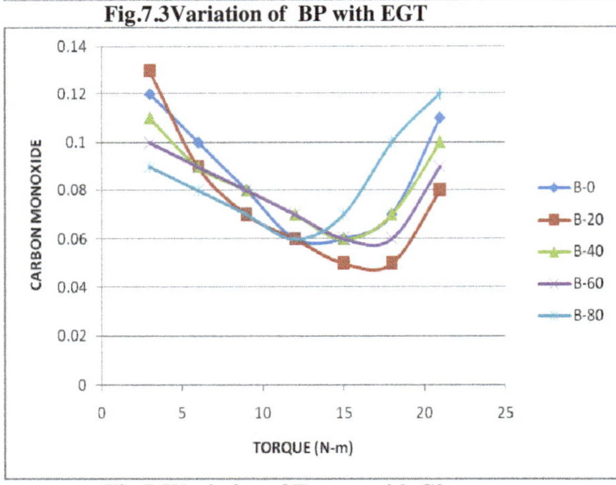

Fig.7.5Variation of Torque with C0

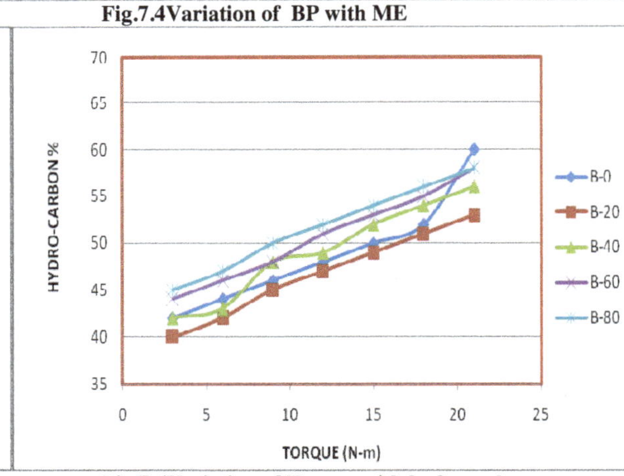

Fig.7.6Variation of Torque with hydrocarbon

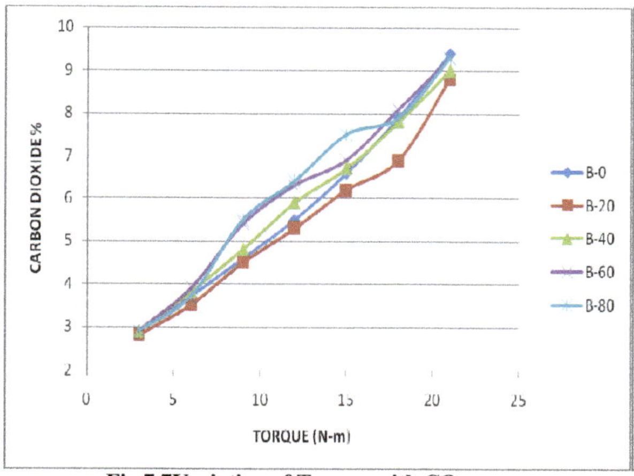

Fig.7.7Variation of Torque with CO_2

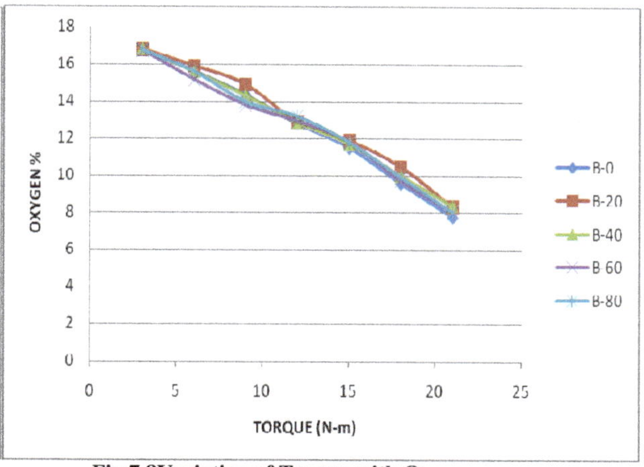

Fig.7.8Variation of Torque with O_2

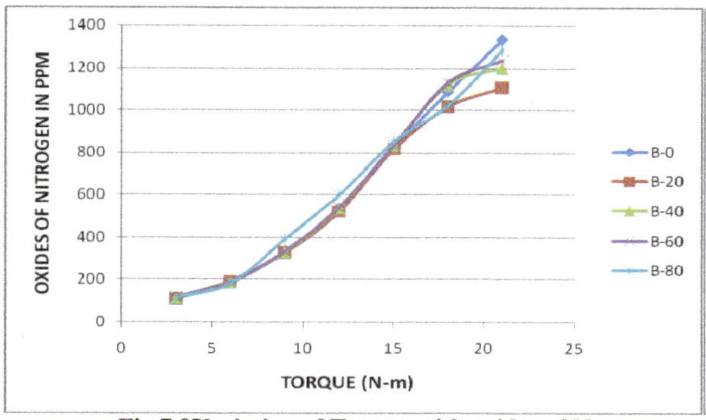

Fig.7.9Variation of Torque with oxides of N_2

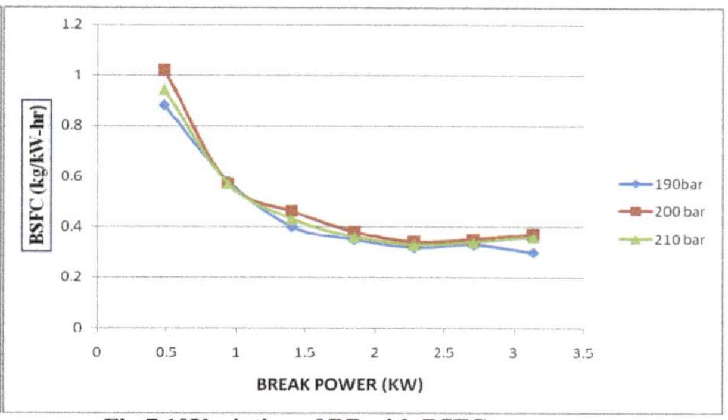

Fig.7.10Variation of BP with BSFC

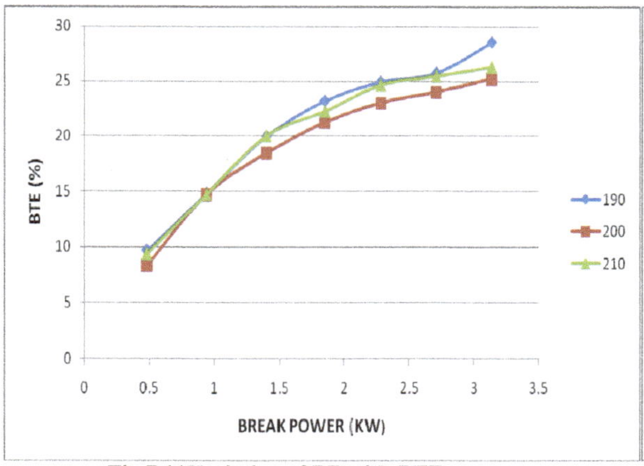

Fig.7.11Variation of BP with BTE

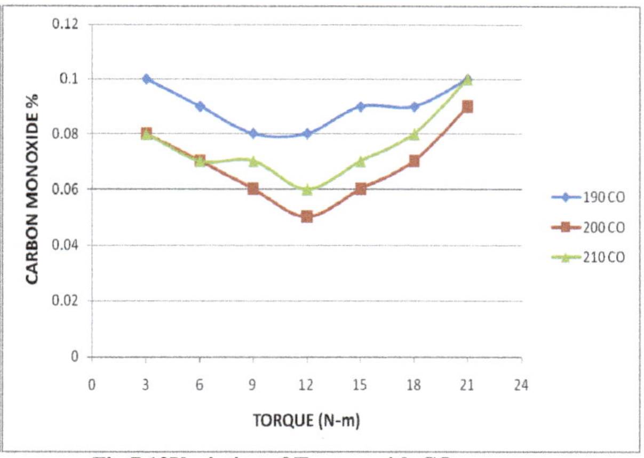

Fig.7.12Variation of Torque with CO

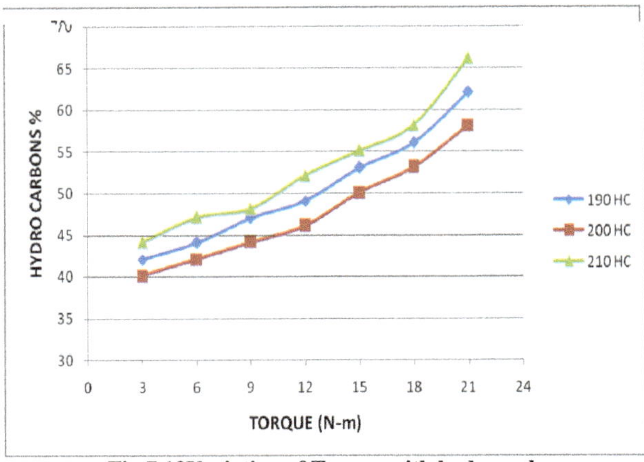

Fig.7.13Variation of Torque with hydrocarbon

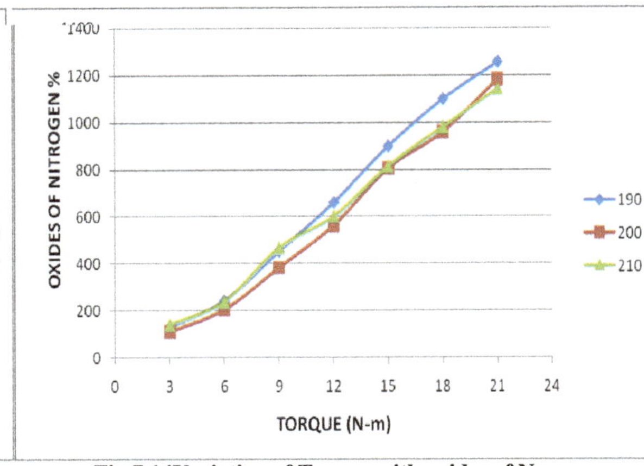

Fig.7.14Variation of Torque with oxides of N_2

The graph 6.1 shows that BTE increases with the increase in the load however for blends B40, B60, B80, it is observed that BTE is less than the diesel, this is due to lower calorific value of blends. However B20 has got higher value of BTE than that of diesel. Hence it is an optimum blend.

Brake Specific Fuel Consumption

The BSFC of an engine for different blends of RBO is shown in above figure 6.2. Generally it is found that with the increase in the concentration of B80 BSFC value increases and sharply decreases with the increase in the BP for all the blends. The main reasons for this could be that the percentage increase in the fuel required to operate the engine is less than the percentage increase in the brake power due to relatively less portion of heat losses at higher loads. BSFC was calculated on weight basis. Thus higher density resulted in higher value of BSFC. As density of biodiesel was higher than that diesel which means the same fuel consumption on volume basis resulted in higher BSFC in case of 100% biodiesel. The higher mass injection will occur for same volume at same injection pressure due to higher density of biodiesel or blends of RBO.

Exhaust Gas Temperature

The variations of EGT with respect to BP are represented in above figure6.3. In general, the EGT increases with increase in engine loading for all the fuel tested. The mean temperature increased linearly from 174^0C at no load to 430^0C at full load condition. This increase in EGT with load is obvious from the simple fact that more amount of fuel was required to the engine to generate that extra power needed to take up the additional loading. The exhaust gas temperature was found to increase with the increasing concentration of biodiesel in the blends.

Mechanical Efficiency

The variation of mechanical efficiency ($\eta_{mech.}$) with brake power for diesel fuel and different biodiesel blends is shown in the above figure6.4. The mechanical efficiency increases with increase in BP for all fuel modes. Mechanical efficiency of the different blends B0, B40, B60 and B80 were found to be lower than that of the B20.The maximum efficiency obtained is 71.25% for B20 at maximum load.

Emission Characteristics

Carbon monoxide Emissions

Variation of CO emissions with engine loading for different fuel is compared in Figure6.5.

The minimum CO produced was found in B80 and it was observed that a reduction of 50%, as compared to diesel. Also it is observed that the CO emissions for biodiesel and its blends are lower than for diesel fuel.

Hydrocarbon Emissions

Above figure6.6 shows the emission characteristics of hydro-carbon v/s load. The reduction in HC was linear with the addition of biodiesel for the blends tested; these reductions indicate a more complete combustion of the fuel. Hydro-carbon emission is more in B0 as compared with the diesel and other blends. Hydro-carbon emission is less in the B20 blend compared with all other blends.

Carbon dioxide Emissions

CO_2 emission increases as the load increases, the maximum CO_2 emission was found in B0. From the above figure6.7 emission of $C0_2$ for different value of blends is near with value of the diesel.

Oxygen Emissions

For methyl ester and its blends, the graph indicated that the O_2 level is comparatively higher in all blends compared to diesel. At all load conditions, B80 shows increasing trend with diesel fuel. Level O_2 of for blends of was slightly in increasing order as blend ratio increased; this may be due to the fact that fuels are oxygenated. The higher O_2 level in fuel blends is always preferred

Nitrogen oxides Emissions

From the above figure6.9 NOx emission for diesel and all the blends increases with increase in loads.

The nitrogen oxides emissions formed in an engine are highly dependent on combustion temperature, along with the concentration of oxygen present in combustion products.

In general, the NOx concentration varies linearly with the load of the engine. As the load increases, the overall fuel- air ratio increases, resulting in an increase in the average gas temperature in the combustion chamber, and hence NOx formation, which is sensitive to temperature increase.

For different injection pressure (IP)

Tests were conducted at three different injection pressures (190, 200 and 210bars) at constant speed of 1500 rpm and at full load condition (3.72 kW) for diesel, B20 blend.

PERFORMANCE CHARACTERISTICS

Brake Specific Fuel Consumption

The above fig 6.10 shows the variation of brake specific fuel consumption with the BP. We observed from the above figure brake specific fuel consumption initially at higher level and with increasing in the load brake specific fuel consumption decreases. In the above three injector pressures 190 bar has higher fuel consumption

Brake Thermal Efficiency

The above figure 6.11 shows the variation of brake thermal efficiency with the BP. We observed from the above figure the brake thermal efficiency gradually increases with the BP. In the above three injector pressures maximum thermal efficiency was obtained at a pressure of 190 bar.

Emission Characteristics

1) Carbon monoxide Emissions

Carbon monoxide emissions from a diesel engine mainly depend upon the physical and chemical properties of the fuel. The bio diesel itself contains 11% of oxygen which helps for complete combustion. From Figure 6.12 it is found that the amount of CO increases at part loads and again greater increase at full load condition for biodiesel. The carbon monoxide emission increases when fuel air-ratio becomes grater. The CO emission for fuels used at full injection pressure is approximately 32% lower than the diesel. The lowest CO emission was observed at 200 bar is 0.32% for biodiesel

Hydrocarbon Emissions

The Hydrocarbon increases with increase in injection pressure for rice bran bio diesel. It is observed from Figure6.13, that the Hydrocarbon emissions for rice bran bio diesel are lower than the diesel fuel, indicating that the heavier hydrocarbon particles that are present in diesel fuel increase Hydrocarbon emissions. The Hydrocarbon emission of rice bran bio diesel the presence of oxygen in the fuel was thought to promote complete combustion that leads to lowering the Hydrocarbon emissions. At 200 bar IP there is minimum Hydrocarbon emissions. An increase in Hydrocarbon which may be due to finer spray, which reduces momentum of the droplets resulting in less complete combustion.

Carbon dioxide Emissions

CO_2 emission increases linearly as the load increases, the maximum CO_2 emission was found at injection pressure of 210bar because of complete combustion of fuel. The CO_2 emissions of all other IP were also higher than the conventional fossil diesel.

NO_X Emissions

Variation of NOx with injection pressure is shown in Figure6.15. The nitrogen oxides results from the oxidation of atmospheric nitrogen at high temperature inside the combustion chamber of an engine rather than resulting from a contaminant present in the fuel. Figures show that the amount of NOx is decreased with increase in injection pressure for bio diesel, this is due to increase in temperature in combustion chamber, as NOx formation is a strong temperature dependent phenomenon. The average NOx emission in case of conditioned bio diesel is 1118 ppm for rice bran bio diesel which is slightly higher than the diesel fuel (1038ppm). NOx emissions were lower at 210 bar injection pressure indicating that effective combustion was taking place during the early part of expansion stroke.

8. Conclusion

Based on the results of this study, the following specific conclusions were drawn:

1. Brake specific fuel consumption for B20 is lower than the diesel fuel and it increases as blend ratio increase.
2. The maximum thermal efficiency for B20 (25.02%) was higher than that of diesel. The brake thermal efficiency obtained for B40, B60, B80 were less than that of diesel.
3. The exhaust temperature increases as a function of the concentration of biodiesel blend i.e. higher the percentage of blend.
4. The fuel properties of rice bran biodiesel except calorific value, all other properties of RBO found to be higher compared to diesel.
5. Viscosity of Biodiesel is higher than that of any other blend & as its concentration increases in the blend, the viscosity of blend increases.

6. From engine performance it is found that the "B20" is optimum blend as it shows higher BTE and lower BSFC & EGT than any other blend.

9. Nomenclature

CV	Calorific value (kj/kg)	NOX	Oxides of nitrogen
CO	Carbon monoxide	BP	Break power (kw)
CO2	Carbon dioxide	FFA	free fatty acid
HC	Hydrocarbon	IP	Indicated power (kw)
02	Oxygen	BSFC	Break specific fuel consumption (kj/kw-hr)
EGT	Exhaust gas temperature (t 0c)	RBO	Rice bran oil
BTE	Break thermal efficiency (%)	B0	100% diesel
ME	Mechanical efficiency (%)	B20	80% diesel, 20%rbo
BD	Biodiesel	B40	60% diesel, 40%rbo
D	Diesel	B60	40% diesel, 60%rbo
B80	20% diesel, 80%rbo		

References

1. L. Canoira, R. Alcantara, J. Carrasco. "Biodiesel from Jojoba oil-wax: Transesterification with methanol and properties as a fuel". Biomass and Bioenergy30 (2006) pp. 76-81.
2. Freedman, B.; Butterfield, R.O.; Pryde, E.H. J. Am. Oil Chem. Soc. 1986, 63, 1375
3. Freedman.B,Pryde EH, Mountus TL "variables affecting the yield of fatty acid esters from vegetable oils". JAOCS 1984;61(10):1638-43.
4. M.M.K Kandasamy and M. Thangavelu. "Investigation on the performance of diesel engine using various biofuels and the effect of temperature variation". Journal of SustainableDevelopment, Volume 2, NO.3, November 2009.
5. T.Balusamy and R. Marappan. 'Performance evaluation of direct injection diesel engine with blends of Thevetiaperuviana seed oil and diesel". Journal of Scientific and IndustrialResearch, Volume 66, December 2007, pp. 1035-1040
6 Sukumar "Performance and emission study of Mahua oil (Madhucaindica oil) ethyl ester in a 4- stroke natural aspirated direct injection diesel engine", Elsevier, Renewable Energy 30, 1269-1278 2005.

BIOGRAPHIES

[1]AnandKumar S Malipatil working at VTU as Assistant Professor .I have a teaching experience of 6 Years and I,am persuing my Phd in heat transfer and fluid flow area.

[2]Manjunath S Bandi , Completed bachelors degree in Mechanical engg. and PG at VTU in Thermal Power Engg. and interested in alternative Fuels.

8

FATIGUE ANALYSIS OF COATED I.C. ENGINE CRANKSHAFT

M. Senthil kumar[1], Dr.S.Ragunathan[2], M.Suresh[3]

[1] Asst.Professor, Dept. of Mechanical Engg.,Sona College of Technology, Salem.
[2] Principal, Jayalakshmi Institute of Technology, Thoppur.
[3]M.E, Engineering Design, Sona College of Technology, Salem.
msksona@gmail.com
ragusubramanian@gmail.com
sureshmsona@gmail.com

Abstract

A survey was conducted on crank shaft used in heavy vehicles. It was reported that the failure has been identified at crankpin bearing location and in the pin contact region with the crank web. Mechanical and metallurgical properties of the crank shaft including chemical composition, metal structure were estimated through optical emission spectroscopy and EDAX with SEM images. The crankshaft model was designed through Pro-E with exact dimensions. Fatigue analysis of modeled crankshaft is done with ANSYS 14 workbench. Nickel and Alumina coating were provided over crank pin region of modeled crank shaft and their strength is estimated as solution to identified fatigue failure. The strength of the crankshaft before coating and after coating were compared and analyzed.

Keywords-crankshaft, failure analysis, coating, finite element analysis, fatigue.

1. Introduction

Crankshaft is the heart of an Internal Combustion Engine [1]. The reciprocating motion of Piston is converted into rotary motion by crankshaft. Crankshafts are generally subjected to torsional stress and bending stress due to self-weight or weights of components or possible misalignment between journal bearings. Crankshaft failures may be resulted from by several causes which are oil absence, defective lubrication on journals, high operating oil temperature, misalignments, improper journal bearings or improper clearance between journals and bearings, vibration, high stress concentrations, improper grinding, high surface roughness, and straightening operations. The crankshaft faults caused high cost of maintenance in automotive industry [2].

Farzin H et al [3], have conducted a dynamic simulation on a crankshaft of a single cylinder four stroke engine. Finite element analysis was performed to obtain the variation of stress magnitude at critical locations. The pressure-volume diagram was used to calculate the load boundary condition in dynamic simulation model, and other simulation inputs were taken from the engine specification chart. The dynamic analysis was done analytically and was verified by simulation in ADAMS which resulted in the load spectrum applied to crank pin bearing. This load was applied to the FE model in ABAQUS, and boundary conditions were applied according to the engine mounting conditions. The analysis was done for different engine speeds and as a result critical engine speed and critical region on the crankshaft were obtained. Stress variation over the engine cycle and the effect of torsional load in the analysis were investigated. Results from FE analysis were verified by strain gauges attached to several locations on the crankshaft. Results achieved from aforementioned analysis can be used in fatigue life calculation and optimization of this component.

S.M.Sorte et al [4] , have analysed the stress and design optimization of a single cylinder crankpin of TVS Scooty Pep crankshaft assembly. Three-dimensional models of crankshaft and crankpin were created using Pro/ENGINEER software and ANSYS was used to analyze the stress status on the crankpin. The maximum deformation, maximum stress point and dangerous areas are found by the stress analysis. They concluded analytically that the value of Max.shear stress is 112N/mm^2 and Max. Bending stress is 180 N/mm^2 which is more than the value of allowable stress value of material without taking any factor of safety. The relationship between the crank rotation and force acting on crank pin would provide a valuable theoretical foundation for the material selection and design optimization of crankpin and engine design.

R.M. Metkar et al [5], have published a review paper and it considers crankshaft as an important part of engine components, as forces which are acting on crankshaft are many and variable in nature. This paper presents an idea about research undertaken or completed on fatigue life of crankshaft, methods available and role of Finite Element Analysis (FEA) and experimental techniques used.

Jian Meng et al [6],have done stress analysis and model analysis of four cylinder engine crankshaft using FEM. The three dimensional model of diesel engine crankshaft was created by Pro-e and also they analysed the vibration modal , the distortion and stress status of crank throw and they found the dangerous areas by stress analysis. The relationship between the frequency and the vibration modal was examined by the modal and harmonic analysis of crankshaft using ANSYS. They concluded that the maximum deformation appeared at the centre of crankpin neck surface. The maximum stress appeared at the fillets between the crankshaft journal and crank cheeks, and near the central point journal. The edge of main journal was high stress area. The failure was due to bending fatigue.

Jamin Brahmbhatt et al [7], have conducted dynamic simulation of single cylinder four stroke diesel engine crankshafts. The three dimensional model of diesel engine crankshaft was created using solid works software. They analyzed the variation of stress magnitude at critical location by using FEM software. Simulation inputs were taken from the engine specification chart. The dynamic analysis was done using FEA software which resulted in the load spectrum applied to crank pin bearing. This load was applied to the FE model in ANSYS and boundary conditions were applied according to the engine mounting conditions. Stress variation over the engine cycle and the effect of torsion and bending load in the analysis were investigated. Von-misses stress was calculated theoretically and FEA. The relationship between the frequency and the vibration modal was examined by the modal and harmonic analysis of crankshaft using ANSYS.

Ali Keskin et al [8], have gone for failure analysis of nodular graphite cast iron crankshaft used in petrol engine. They tested mechanical and metallurgical properties of the crankshaft including chemical composition, micro-hardness, tensile properties and roughness and were compared with the specified properties of the crankshaft materials. In the comparison, there were no metallurgical defects apart from slightly higher carbon content. All other measured values were within the specified values. The reason identified for the failure was the thermal fatigue because of contact of journal and bearing surface. This condition led to the formation and growth of fatigue cracks. The contact was resulted from defective lubrication or high operating oil temperature. S.Bhagya Lakshmi et al [9], have designed the crankshaft of Honda engine and assembled by connecting rod in Pro-e. The designed model of crankshaft was analysed using Pro-e by using its mechanism. The piston acceleration was calculated theoretically and compared with experimental result. The peak pressure acting on the engine crankshaft was 63 bar. The crank angle was calculated with respect to seconds as it takes 0.008 sec for each stroke with two revolutions. This value was calculated using spread sheet. It became base for dynamic analysis using Pro-E.

2 .FAILURE ANALYSIS

Mechanical and metallurgical properties of the crankshaft including chemical composition, and microstructure were studied to understand the process of failure.

A.CHEMICAL ANALYSIS OF CRANKSHAFT

The test result from the OES – Foundry master showed that the material used in the crank shaft is **EN9 (070M55** or **AISI/SAE 1055).**

Fig.1. Portion of Chemical composition measured

COMPOSITION	RANGE	MEASURED
Carbon	0.50 - 0.60%	0.52%
Manganese	0.50 - 0.80 %	0.76%
Silicon	0.05 – 0.35 %	0.25%
Sulphur	0.00 – 0.06 %	0.025%
Phosphorous	0.00 – 0.06 %	0.017%

Table .1. Chemical Composition of crankshaft

PROPERTIES	VALUES
Young's Modulus	190-210GPa
Density	7850Kg/m^3
Poisson ratio	0.27-0.3
Tensile stress	700-800 Mpa
Yield stress	355-500 Mpa
Hardness	201-302 HB

Table.2 Material Properties of crankshaft

B.SEM IMAGE (SCANNING ELECTRON MICROSCOPE)

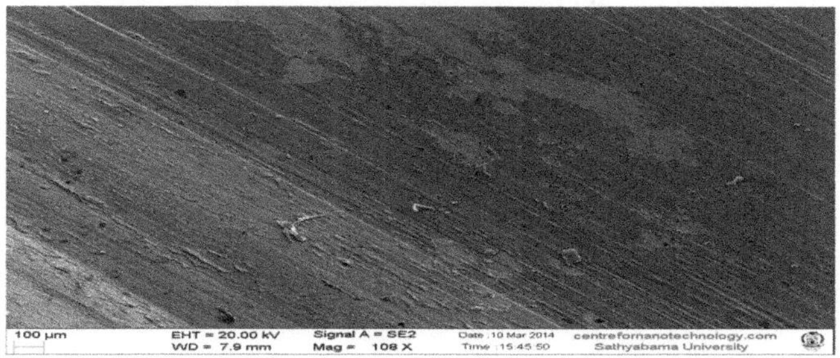

Fig.2 SEM Image With Magnification 108x

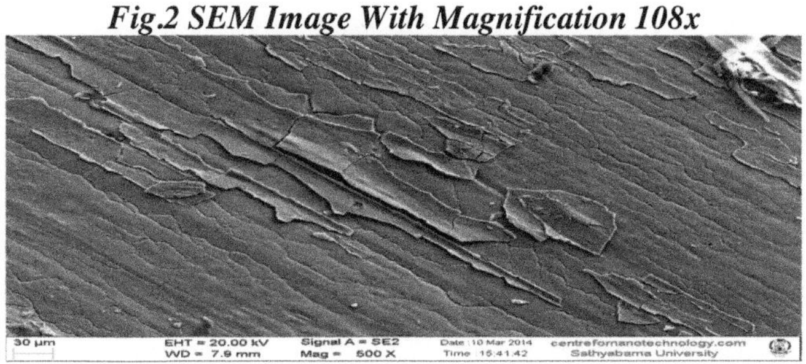

Fig.3. SEM Image with Magnification 500x

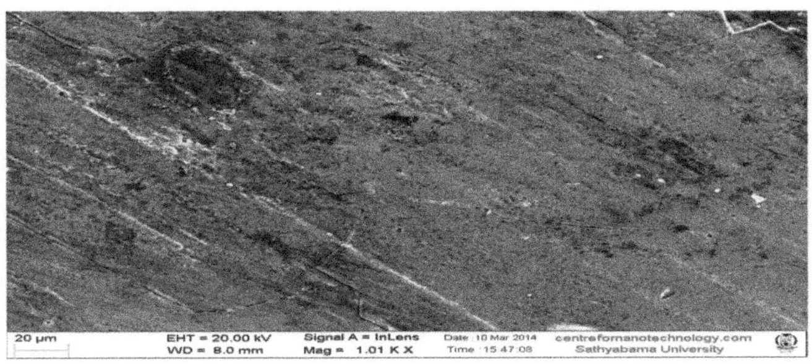

Fig.4.SEM Image with Magnification 1.01k X

Fig.5. SEM Image with Magnification 20.00 KX

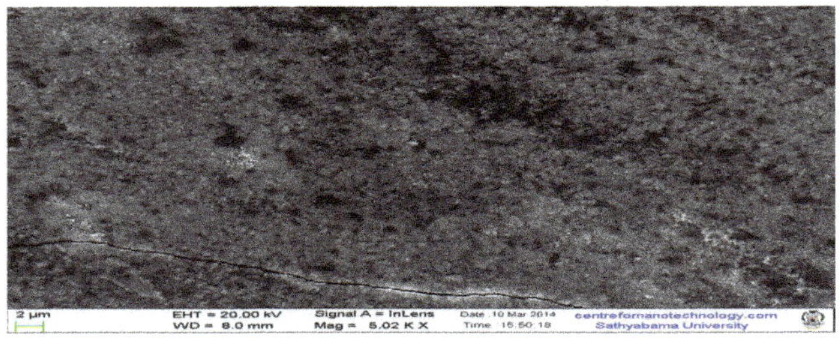

Fig.6. SEM Image with Magnification 500 X

The scanning electron microscope (SEM) uses a focused beam of high-energy electrons to generate a variety of signals at the surface of solid specimens. The signals that derive from electron-sample interactions reveal information about the sample including external morphology, chemical composition, and crystalline structure and orientation of materials making up the sample.
SEM image shows chemical composition and crystalline structure of crankshaft materials at various magnifications. The material structure obtain from the SEM image is martensite.

C.ENERGY DISPERSIVE X-RAY ANALYSIS (EDX) COMPOSITION ANALYSE EDAX REPORT

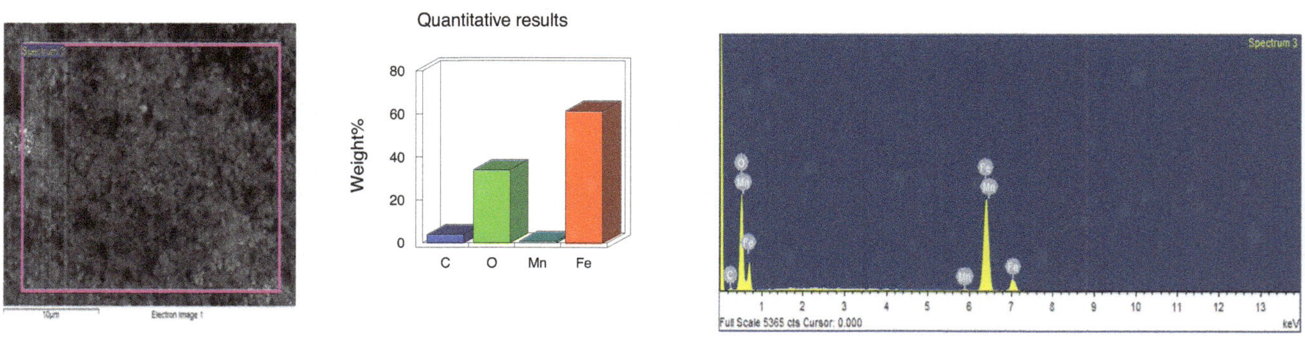

Fig.7. Energy Dispersive x-ray analysis image

The material structure and compositions are found using energy dispersive x-ray analysis. The structure formed is martensite. The chemical composition test results are given below

ELEMENT	WEIGHT%	ATOMIC%
C K	3.78	8.83
O K	34.09	59.89
Mn K	0.71	0.36
Fe K	61.42	30.91
Totals	100.00	100.00

***Table.3*. chemical composition**

3. DESIGN CALCULATIONS

A.ENGINE SPECIFICATION

DESCRIPTIONS	DETAILS
Engine Type	4Stroke, 6 Cylinder, Air Cooled
Displacement	97.2 cc
Bore and Stroke	104X 113mm
Compression Ratio	17:1
Max. Power	97 kw @ 2400rpm
Max. Torque	550 kgm @ 1600rpm
Transmission	6 Speed, Syncro mesh

Table.4. Engine Specification

B.FORCE ANALYSIS OF 4 STROKE 6 CYLINDER DIESEL ENGINE

Volume (V) = π.D2.L / 4

= π / 4x0.104^2 x 0.113

= 9.599x10-4 m3

Mean Effective Pressure (Pmean)

= P x 60 x n / N x V

= 97x60x2/2400x9.599x10-4

= 5.05 Mpa

Piston Load (FL)= Pmean x Area

= 5.05x8.0154x10^3

FL = 43 KN

Crank effort (FT)

θ	TIME	Inertia Force (F_i) in KN	Piston Effort (F_P) in KN	Ø	Crank Effort (F_T) in KN	Firing Order
135	0	3.533	46.546	12.972	25.331	1
139		3.550	46.563	12.021	23.065	
140	1	3.552	46.565	11.774	22.496	5
143		3.553	46.566	11.014	20.786	
145	2	3.550	46.564	10.491	19.644	3
150		3.533	46.547	9.133	16.793	
155	3	3.508	45.652	7.710	13.952	6
160		3.480	46.494	6.233	11.130	
165	4	3.452	46.466	4.713	8.326	2
170		3.430	46.444	3.160	5.539	
171	5	3.426	46.440	2.846	4.984	4
175		3.415	46.423	1.585	2.766	

Table.5. Details of crank effort

Where

Crank effort (F_T) = F_P. Sin (θ + $\varnothing$) / Cos$\varnothing$

Piston Effort (F_P)= $F_L + F_i + W_R$

Maximum force on piston (F_L) = 43 KN

Inertia force (F_i) =-m_R x rω^2(Cos θ + (Cos2 θ / n))

Angular speed (ω)= 251.33 rad/sec

Crank Radius (r) = 0.565 m

Connecting rod length (l) = 0.178 m

n = l/r

= 3.15

4. FINITE ELEMENT ANALYSIS (FATIGUE ANALYSIS)

Finite element analysis is a computer based numerical technique for calculating the strength and behavior of engineering structures. Finite element analysis is a technique to simulate loading condition on a design and determine the design's response to those conditions. We used Pro-E as design software for modeling and analyzed using ANSYS- 14 work bench.

A.MODELING

The dimensions of the crankshaft is measured accurately (reverse engineering). By using these dimensions, model of the crank shaft was drawn using Pro-E software

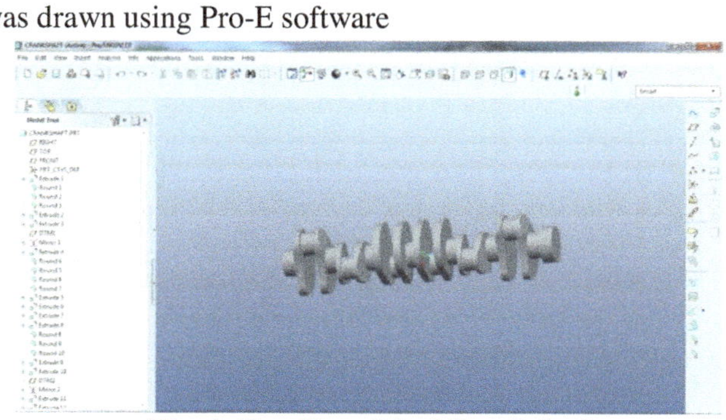

Fig .8. Pro-E model of Crankshaft

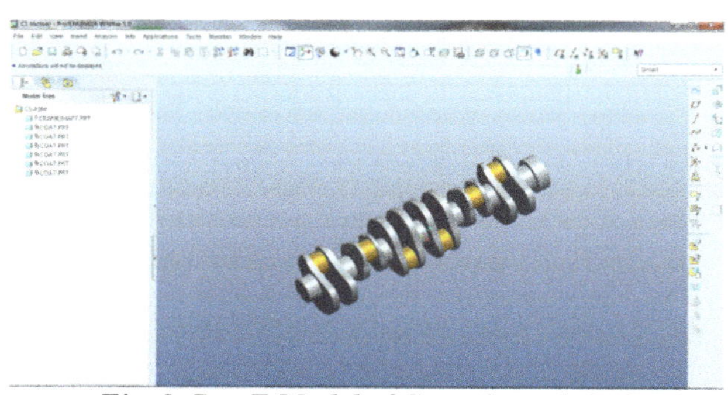

Fig. 9. Pro-E Model of Coated crankshaft

B.ANALYSISOF CRANK SHAFT

We used ANSYS-14 (WORK BENCH) software for analysis. The model was imported to ANSYS 14 work bench. Material property and S-N curve details are entered for the crankshaft (EN9).By doing reverse engineering process the crankshaft was measured and drawn in Pro E and imported to the Ansys 14 work bench. From fatigue analysis we obtained the result like deformation, equivalent stress, alternating stress, life and damage of crankshaft.

Meshing was generated for coated crankshaft. Meshing is used to convert the single component into number of elements and nodes. This process increases the accuracy of result and can give us the stress and deformation value at any desired location.

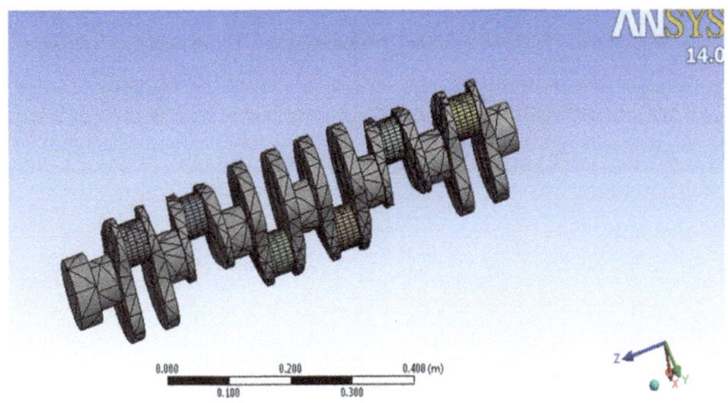

Fig.10. Meshed geometry of coated crankshaft

Then boundary conditions were applied to the crankshaft.

- Arresting the bearing supports on both sides.
- Applying gas pressure on the top of the crank pin portion depends up on the firing order and angle of crank. In the ANSYS picture point A shows fixed support and other points B,C,D,E,F,G shows force acting on various crank pin (B F1:25000N,C F5:22000N,D F3:19000N,E F6:13000,F F2:8000,G F4:4000)

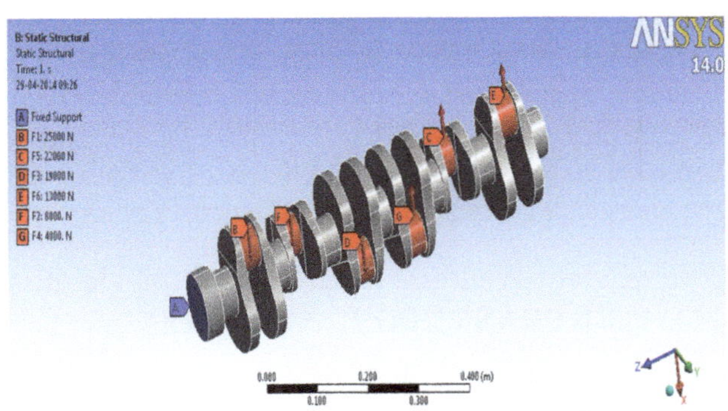

Fig.11. Boundary condition applied to coated crankshaft

B.1.ANALYSIS OF CRANK SHAFT WITHOUT COATING

The value of von-mises stress spread across the crank shaft are indicated in colour bandwidth which has Red colour indicates maximum stress value as 1.0696X10^8 Pa.

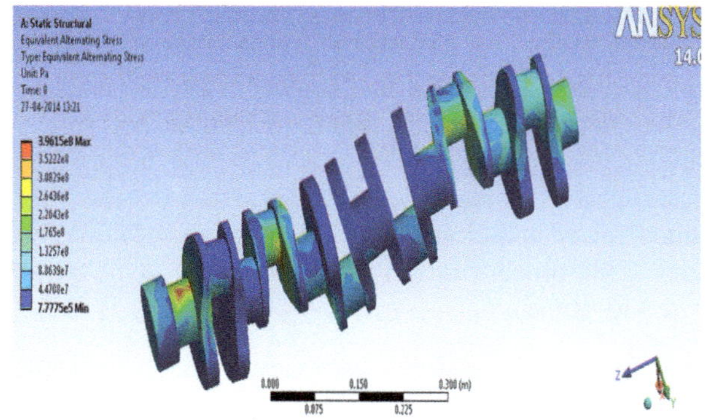

Fig.12. Equivalent Alternating Stress Of Non-Coated Crankshaft

The stress is maximum at the fillet region of crankpin which is shown in RED colour. The maximum value of alternating stress is 3.9615e8 Pa.

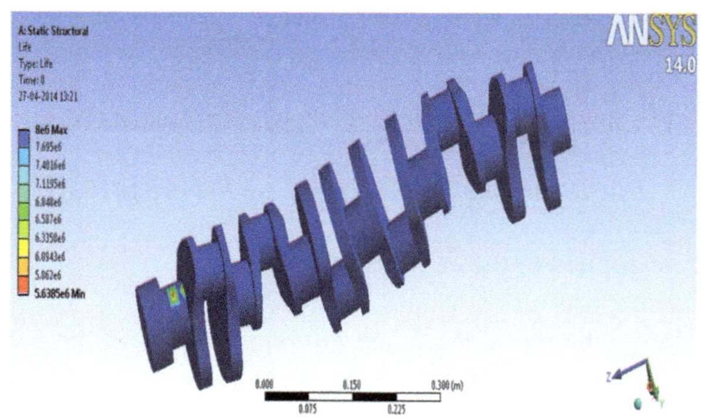

Fig.13. Life of Non-Coated Crank Shaft

The maximum life of crankshaft is 8e6 cycles and the minimum value is 5.6385e6 cycles.

Poisson ratio and density were given to material models. By doing fatigue analysis, we found that the life of crankshaft is finite and it is 5.63 x 10^6 cycles. In order to increase the life of the crankshaft, we have given coating of Al_2O_3 and Ni to the crankshaft at crankpin portion. The material property of Alumina and nickel was given.

B.2.ANALYSIS OF Al_2O_3 COATED CRANKSHAFT

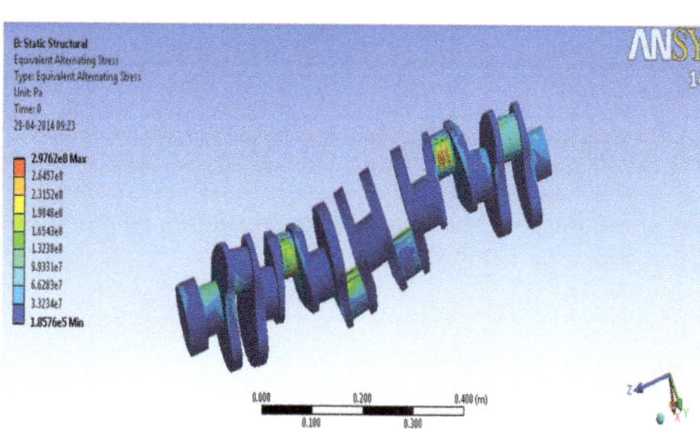

Fig.14.* Equivalent Alternating Stress *of Al_2O_3 coated crankshaft

The red color shows that the Maximum Alternating stress value as 2.9762e8 N/m^2 and the blue color shows that the Minimum Alternating stress value as 1.8576e5 N/m^2.

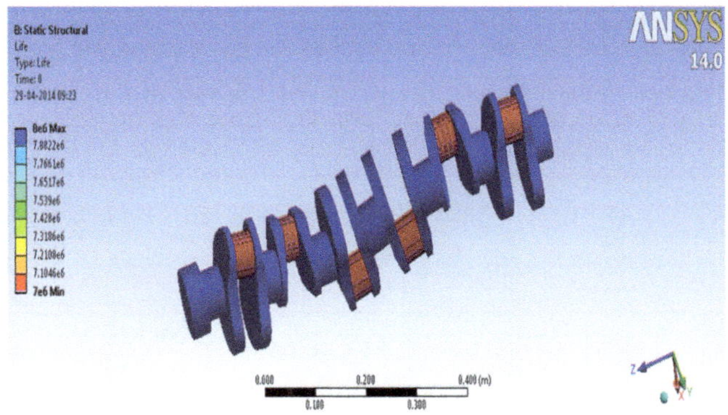

Fig.15.* Life of *Al_2O_3 coated crankshaft

The blue color shows that the Maximum life value as 8e6 cycles and the red color shows that the Minimum life value as 7e6 cycles .

B.3.ANALYSIS OF NICKEL COATED CRANKSHAFT

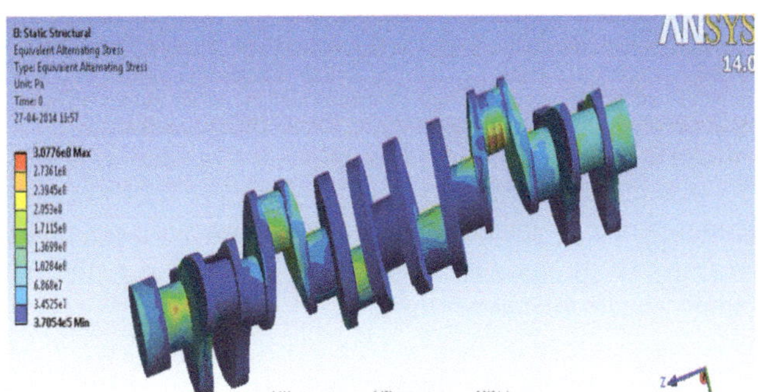

Fig.16.Equivalent Alternating Stress of Ni coated Crankshaft

The red color shows that the Maximum Alternating stress value as 3.0776e8N/m^2 and the blue color shows that the Minimum Alternating stress value as 3.7054e5N/m^2.

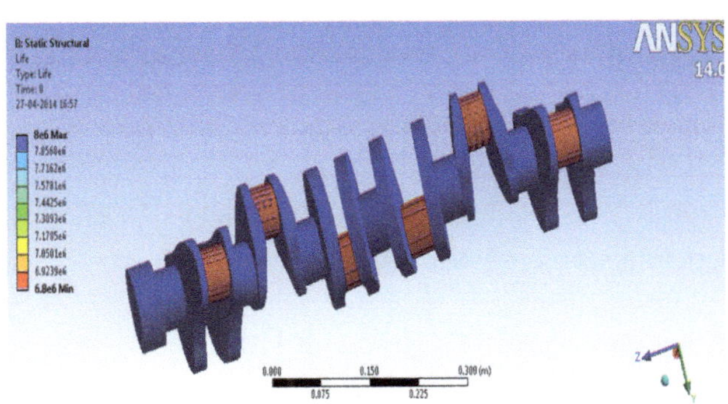

Fig.17. Life of Ni coated Crankshaft

The blue color shows that the Maximum life value as 7.8568e6 cycles and the red color shows that the Minimum life value as 6.8e6 cycles .

5. RESULT AND DISCSSION

We have chosen the heavy vehicle Engine Crankshaft for analysis of fatigue strength. From our analysis, the fatigue life of Non-coated crankshaft is 5.63×10^6 cycles. In order to increase the life of the crankshaft, we have given coating by Al_2O_3 and Nickel. The life of Al_2O_3 Coated Crankshaft is 7×10^6 cycles and Nickel coated Crankshaft is 6.8×10^6 cycles. By comparing the result Coated Crankshafts have better life than Non-coated crankshaft.

	Non-Coated crankshaft	**Al_2O_3Coated Crankshaft**	**Nickel coated Crankshaft**
Fatigue stress N/mm^2	396.15	297.62	307.76
Fatigue Life (cycles)	5.63×10^6	7×10^6	6.8×10^6

Table 6.Comparison of fatigue life of coated and Non-coated crankshaft

6. FUTURE WORK

The coating of Nickel and Alumina on any material is found to be increasing its fatigue life which validates the claim of coating the material improve the fatigue life of crank shaft. Hence, the amount of increase in fatigue life on coated crank shaft is under experimental testing.

REFERENCES

1. Silva.F.S. (2003), "Analysis of a vehicle Crankshaft Failure", Pergamon, Engineering Failure Analysis, 10,605-616.

2. Xue-qin Hou et al (2010), "Fracture Failure Analysis of Ductile Cast Iron Crankshaft in a Vehicle Engine", ASM International.

3. Farzin H . et al (2007)," Dynamic load and stress analysis of a Crankshaft", SAE International Journal, 2007- 01-0258.
4. Sorte. S.M. et al (2013), "Stress Analysis and Design Optimization of Crankpin", International Journal of science and Modern Engineering , Vol 1 , Issue 4 .

5. R.M. Metkar,V.K. Sunnapwar and S.D. Hiwase 2011,"A Fatigue Analysis And Life Estimation of Crankshaft",International Journal of Mechanical and Materials Engineering (IJMME), Vol.6 (2011), No.3, 425-430
6. Jian Meng, (2011), "Finite Element Analysis of 4-Cylinder Diesel Crankshaft",Image Graphics and Signal Processing, 5, 22-29
7. Jamin Brahmbhatt et al (2012),"Design and Analysis of Crankshaft for Single cylinder 4-stroke Diesel Engine", International Journal of Advanced Engineering Research and Studies, Vol 1, Issue 4, 88-90.

8. Ali KESKIN et al (2010), "Crack Analysis of a Gasoline Engine Crankshaft", Gazi University Journal of science, 23 (4): 487-492.

9. BhagyaLakshmi.S, et al (2012),"Dynamic Analysis of Honda Engine Crankshaft", International Journal of Engineering and Innovative Technology Vol.2, Issue1.

10. Hassan Nosrati1, Seyedhadi Tabaiyan, Seyed mohammad mahdihadavi (2012), "Effect of Nickel Pulse Electroplating Parameters on ST12 Steel" ,ISSN: 2090-4215 (154-158).

11. Imran M Quraishi ,Mrs.Madhuvi S Harne (2012), "Stress analysis and optimization of crankshaft under dynamic loading",ISSN 0976 – 6359 Volume 3, Issue 3, pp. 429-437.

12. Alfares. M.A. et al (2007), "Failure Analysis of a Vehicle Engine Crankshaft", ASM International.

13. Becerra.J.A.et al (2011), "Failure Analysis of Reciprocating Compressor Crankshafts", Elsevier, Engineering Failure Analysis 18, 735-746.

9

INFLUENCE OF CORNER GEOMETRY ON THE FLOW STRUCTURE AND FLOW CHARACTERISTICS FOR FLOW PAST A SQUARE CYLINDER AT Re=150

Nidhul K

PG Student, Government College Of Engineering Thrissur, nidhul07@gmail.com

Abstract

Flow past a square cylinder with and without corner modifications is carried out numerically by using CFD fluent. The wake is generated by a uniform flow of Reynolds number (Re) 150 based on the characteristic length of the cylinder, D. 2D unsteady numerical simulation is done using FVM employing pressure based solver and PISO scheme. A computational grid independence study has been done to obtain a grid resolution which predicts the results without any discrepancies. The flow separation point for the square cylinder with and without corner modifications is obtained. The pressure distribution in the near wake region and around the square cylinder surface is also investigated for different corner geometries. The results are presented in the form of coefficient of pressure Vs Domain length/D, Coefficient of pressure along the cylinder wall and wall shear stress along the cylinder wall. The results indicate that the flow separation in case of square cylinder without sharp corners is delayed. The adverse pressure gradient along the surface of the cylinder and in the near wake region is smaller for round corners. The tangential velocity of square cylinder with sharp corners is large when compared with modified corners.

Keywords: Square Cylinder, Corner modifications, Reynolds number, wake, grid independence, flow separation point, Pressure distribution, Lift coefficient, Drag coefficient.

1. Introduction

Flow past bluff bodies has attracted a great deal of attention in the literature because of its practical significance in engineering e.g., Tall buildings, monuments, and towers are permanently exposed to wind. Bluff bodies are structures with shapes that significantly disturb the flow around them, as opposed to flow around a streamlined body. A bluff body is one in which the length in the flow direction is close to or equal to the length perpendicular to the flow direction. Examples of bluff bodies include circular cylinders, square cylinders and rectangular cylinders.

Flow around a bluff body creates a large region of separated flow and a massive unsteady wake region in the downstream when compared to a streamlined body. Due to the adverse pressure gradient in the flow direction, flow separates and flow reversal occurs which creates rotational lump of fluid masses known as eddies. Eddies formed are detached periodically from either side of structure, known as vortex shedding, generates velocity fluctuations in the wake region. Alternate shedding of vortices also induces a periodic oscillatory force in the transverse direction, known as the lift force. When the natural frequency of structure is close to the shedding frequency of the vortices, the phenomenon of resonance occurs. A lot of research has been carried out on flow past single square cylinder and passive methods to control vortex shedding. However, there has been no complete investigation on the effect of corner modification on the flow characteristics around a square cylinder. Therefore, the simulation of unsteady flow past square cylinder with corner modifications has practical relevance.

Sohankar et al. [1] constructed a non-uniform mesh near the wall of a square cylinder and, at 5 diameters away from the cylinder surface, a uniform mesh was applied. The size of the smallest cell was located at the leading edge of the square cylinder with cell size of h/D = 0.004. Inoue et.al [2] constructed a non-uniform mesh but divided the computational domain into three regions, each with a different grid ratio. The smallest cell was located along the edges of the square cylinder with the value of h/D = 0.01. Mohamed Sukri Mat Ali et.al [3] numerically investigated the sensitivity of the computed flow field to flow parameters for a flow with Reynolds number 150.They constructed computational meshes based on reasonable estimates of cell size and grid stretching ratios. Doolan [4] investigated the grid convergence for three different grid resolutions on DNS around a square cylinder and found that the solution converged when the smallest cell size along the square cylinder edge was h/D = 0.0167. Bearman et al. [5] investigated the corner radius effect on the hydrodynamic forces on cylindrical bluff bodies subjected to an oscillatory flow. They found that the drag coefficient, Cd, was sensitive to the corner radius in a steady flow and even more so in an oscillatory flow. Kawai [6] also has investigated the effects of corner modifications (such as corner cut, recession and rounding) on the flow characteristics and flow pattern past square prisms. He concluded that corner rounding is the best option to provide aerodynamic stability to prisms. Tamura et al· [7] and Tamura and Miyagi [8] investigated numerically and experimentally the aerodynamic forces on square cylinders and observed a decrease in the wake width as well as C_d with the round corners. Dalton and Zheng [9] presented numerical results for a uniform approach flow past square and diamond cylinders, with and without corner modifications at Re=250 and 1,000. They noted that rounding corners of the bluff bodies produced a noticeable decrease in the calculated drag and lift coefficients. B. H. Lakshmana Gowda et al. [10] studied the near wake flow field features of transversely oscillating square section cylinders with different corner radii. Results indicate that increasing the corner radius suppresses the possible instabilities of the cylinder.

Over the years the investigations largely focused on the effect of corner radii on the aerodynamic characteristics, such as drag/lift forces and vortex shedding frequency. However, how the corner variation may alter the pressure variation in the near wake and along the cylinder surface has yet to be sufficiently documented. Hence one objective of this work is to characterize quantitatively the corner effects on the near-wake flow characteristics and flow structure, complementing the data in the literature.

2. Numerical Simulation Procedure

2.1 Description of model

A cylinder of square cross-section with a dimension of 4cm width and 4cm length is the model taken. R/D ratio for rounded corner is 0.125. For inclined corner, angle of inclination is 45^0. Rectangular corner is made by removing square portion of length 0.005m from each corner as shown in Figure 1.

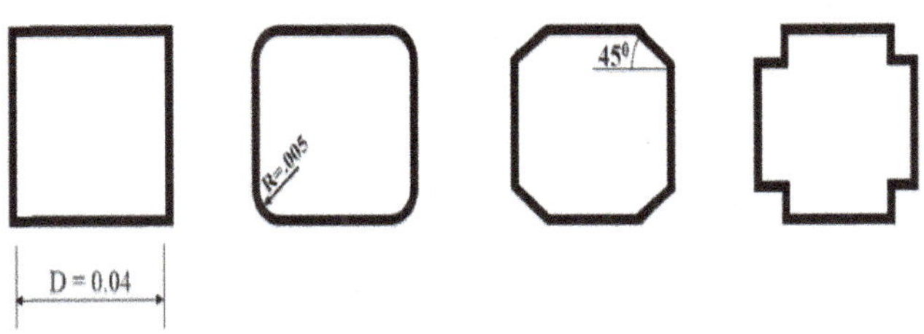

Figure 1: Cylinder with square cross-section and corner modifications

2.2 Flow field formulation

The governing equations on the flow field are the continuity and momentum equations (Navier–Stokes equations), which can be written as follows:

$$\frac{\partial \rho}{\partial t} + div(\rho V) = 0$$

$$\frac{\partial}{\partial t}(\rho u) + div(\rho V u) = -\frac{\partial p}{\partial x} + div(\mu\, grad u) + B_x$$

$$\frac{\partial}{\partial t}(\rho v) + div(\rho V v) = -\frac{\partial p}{\partial y} + div(\mu\, grad v) + B_y$$

Where ρ is the fluid density, μ is the fluid viscosity, V is the velocity vector of the flow field, p is the pressure, and u and v are the velocity components in the x- and y-directions, respectively.
B_x and B_y are also the body forces per unit volume, which are negligible in the present study. The fluid is assumed to be incompressible, and its properties has been taken as ρ=1.225 kg/m^3 and μ=1.7894×10^{-5} kg/m s.

A rectangular domain was used with a length of 38D and a width of 20D, where D=0.04 m is the cylinder diameter. The cylinder center has the coordinates x=10.5D and y=10.5D as shown in Figure 2.

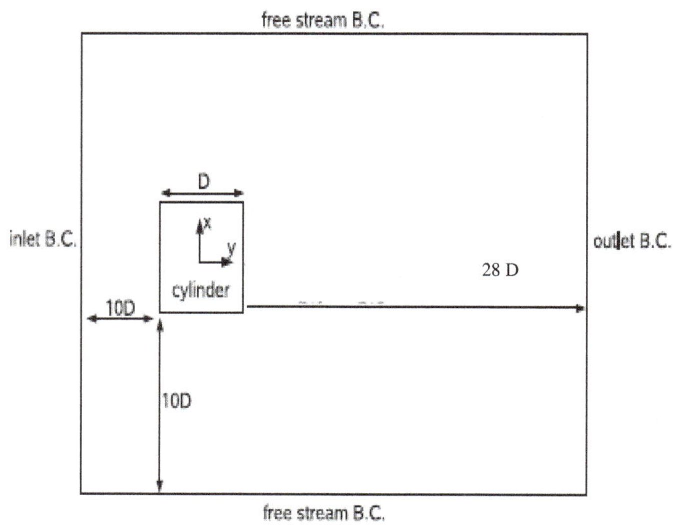

Figure 2: Computational flow domain for flow past a square cylinder

2.3 Discretization method

The governing equations have been discretized using the finite-volume method on a fixed Cartesian-staggered grid with non-uniform grid spacing. The grids in the region of the cylinder boundaries are sufficiently fine in order to achieve the reasonable accuracy. The temporal discretization has been done in conformity with the fully implicit scheme. For the spatial discretization, the hybrid scheme has been employed. PISO (Pressure-implicit with splitting of operators) procedure was applied to calculate the flow field variables.

3. Grid influence

Within the scope of predicting the flow field around a square cylinder using numerical analysis, many similar investigations have been made, but the results always show small discrepancies even though the overall global trends are similar. One of the reasons for these discrepancies is the difference in the construction of the mesh. So grid independence study has been done so that the grid chosen for simulation provides accurate results.
The mean axial velocity over the central plane of domain is shown in Figure 3. Six different 2D grids are used starting with 27200 cells upto 176094 cells.

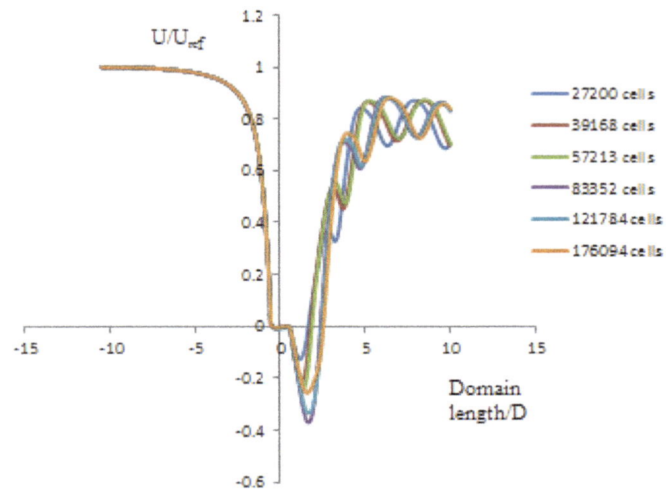

Figure 3: The mean axial velocity distribution on the center plane of the cylinder

The results of this case show a certain dependency on the grid resolution for the grids used. In Figure 3, it is observed that upstream all the runs predict identical results, whereas a faster recovery of mean axial velocity is predicted when the grid resolution is increased but the mean velocity of recirculation is approximately the same in all runs.

4. Results and Discussion

Preliminary simulations have been carried out in order to define adequate computational domains and to assure grid independence. Both uniform and non-uniform meshes have been tested, with different refinements. The values of C_D, C_L and Strouhal number for flow past a square cylinder obtained by simulation are compared with experimental and simulation work of other authors as shown in Table 1.The results obtained indicates that the numerical scheme, domain size, mesh resolution, and other parameters have been adequately chosen to capture the complex wake behaviour.

Then the results i.e. streamlines, vorticity magnitude, flow separation point, tangential velocity, pressure distribution in the near wake region and pressure distribution along the cylinder surface are demonstrated and analysed at Reynolds numbers of 150 for 4 cases i.e. for square cylinder with sharp corners, for square cylinder with round corners, for square cylinder with inclined corners and for square cylinder with rectangular corners.

Table 1: Comparison of results of case study with other authors works

Study	St	C_D	C_L
Experiments(Okajima, 1982; Sohankar et al., 1999)	0.148-0.155	1.40	-
Sohankar et al. (1998)	0.165	1.44	0.230
Doolan (2009)	0.156	1.44	0.296
Inoue et al. (2006)	0.151	1.40	0.40
Ali et al. (2009)	0.160	1.47	0.285
Present study	0.149	1.60	0.44

4.1 Streamlines and Vorticity magnitude contours

For flow over a square cylinder at Re=150, the flow is uniform and symmetrical in the upstream of the cylinder as shown in Figure 4(a). Due to adverse pressure gradient in the flow direction combined with low momentum near the wall surface, the flow streamline detach from the top corner of leading edge. Flow separation and flow reversal occurs. Hence rotating lump of fluid is created in the near wake region. As the flow forms a clockwise eddy, it rushes past the top of the square cylinder faster than the flow at the bottom.

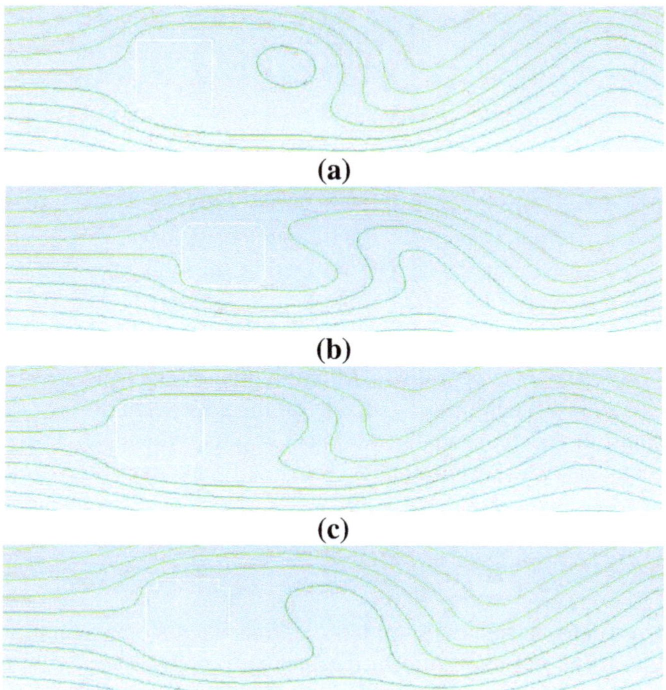

(a)

(b)

(c)

(d)

Figure 4: Streamlines flow around (a) square cylinder (b) square cylinder with round corners (c) Square cylinder with inclined corners and (d) Square cylinder with rectangular corners

This eddy grows in size as they move away from the cylinder upto a certain length from the cylinder trailing edge dissipating energy to smaller eddies downstream. Vorticity magnitude contours for the different cases are shown in Figure 5.

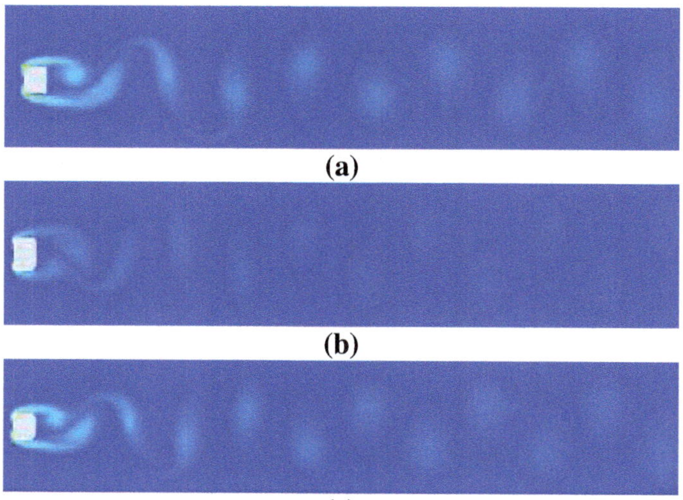

(a)

(b)

(c)

(d)

Figure 5: Vorticity magnitude contours for flow around (a) square cylinder (b) square cylinder with round corners (c) Square cylinder with inclined corners and (d) Square cylinder with rectangular corners

4.2 Tangential velocity and Wake width

Tangential velocity variation for flow over a square cylinder with and without corner modification is shown in Figure 6. It is observed that tangential velocity of square cylinder is large when compared to square cylinder with round, with inclined and with rectangular corners. Due to this separation area for square cylinder with sharp corners is enlarged. Therefore width of the wake behind the square cylinder with corner modification becomes large.

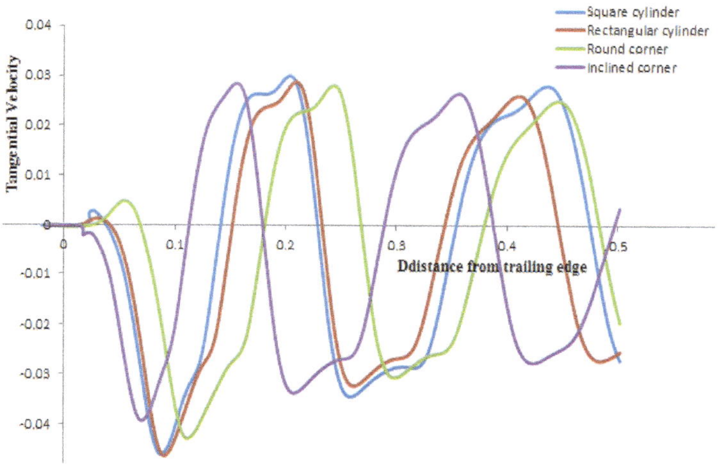

Figure 6: Tangential velocity variation in the near wake region

4.3 Flow Separation

In fluid flow analysis over a bluff body, flow separation is an important term encountered. When the fluid flows over the top surface of square cylinder, flow momentum is quite low due to viscous effects and is thus sensitive to changes of pressure gradient. The flow has to move against adverse pressure gradient in addition to the viscous force. This leads to reduced velocity and wall shear stress. When the flow cannot overcome the adverse pressure gradient in the flow direction, boundary layer separates and flow reversal occurs causing the formation of lump of rotating fluid mass known as eddies.

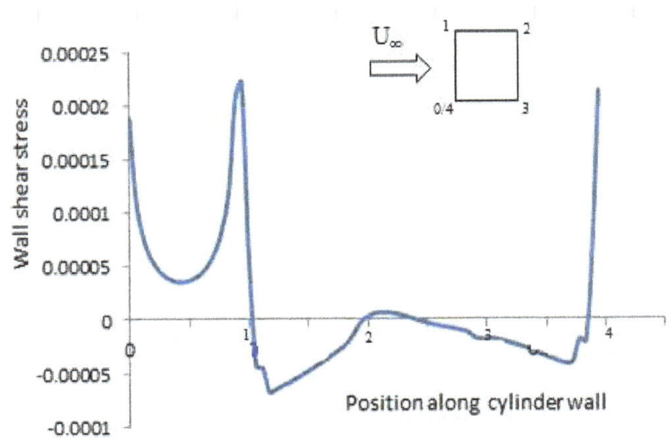

U∞
1
2
0/4
3
0.00025
0.0002
0.00015
0.0001
0.00005
0
-0.00005
-0.0001
Wall shear stress
Position along cylinder wall

(a)

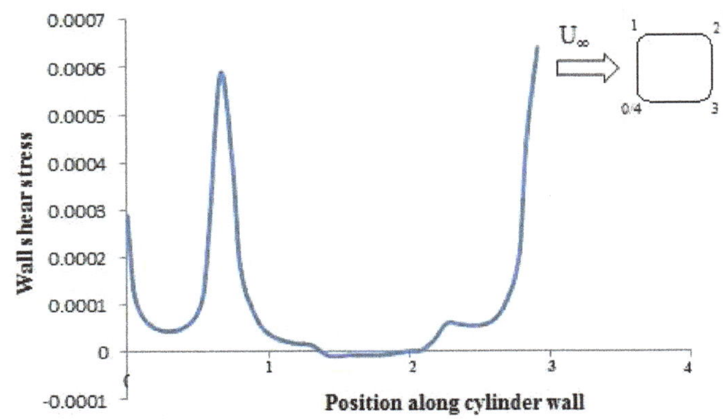

U∞
1
2
0/4
3
0.0007
0.0006
0.0005
0.0004
0.0003
0.0002
0.0001
0
-0.0001
Wall shear stress
Position along cylinder wall

(b)

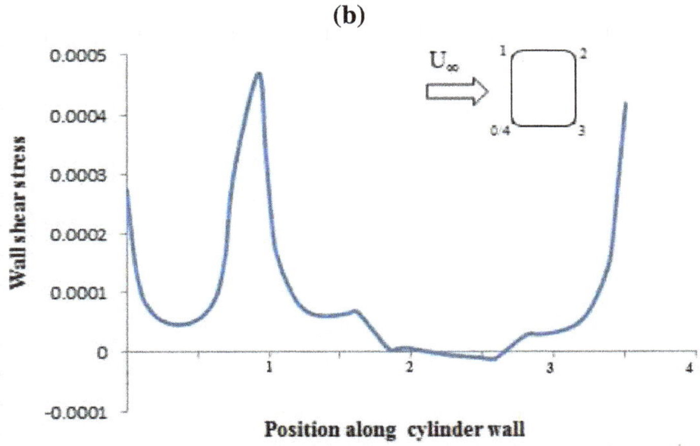

U∞
1
2
0/4
3
0.0005
0.0004
0.0003
0.0002
0.0001
0
-0.0001
Wall shear stress
Position along cylinder wall

(c)

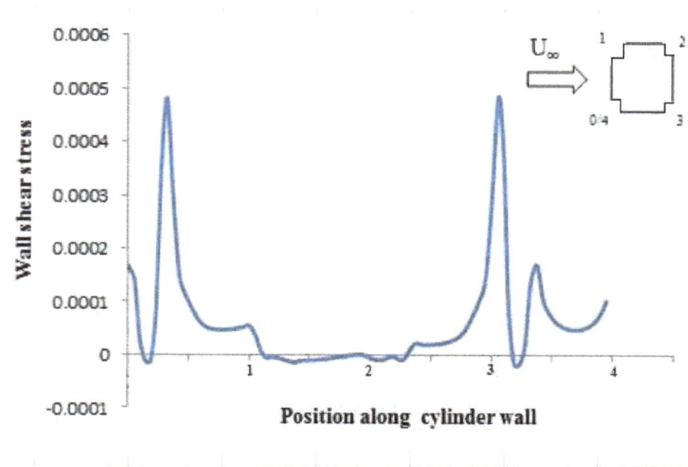

(d)

Figure 7: Flow separation points for flow past (a) Square cylinder with sharp corners (b) Square cylinder with round corners (c) Square cylinder with inclined corners and (d) Square cylinder with rectangular corners

At Re=150, flow separation occurs just after top left corner (1) of the square cylinder with sharp corners where wall shear stress is zero as shown in the Figure 7(a). Whereas with round corners, boundary layer remains attached even after top left corner and separates near the midway of top surface (1-2) of the cylinder as shown in Figure 7(b).For inclined corners, flow separation occurs near top right corner (2) and for rectangular corners, it occurs just after top left corner as shown in Figure 7(c) and Figure 7(d) respectively.

4.4 Pressure Distribution along the wall surface

Apart from velocity profile, pressure distribution is an important parameter in the study of flow around a cylinder. Pressure changes accordingly with the vortices motion in the vicinity of the body. At stagnation point, located at midpoint of face 0-1as shown in Figure 8, flow comes to rest (v=0) and pressure thus reaches maximum value (By Bernoulli's theorem).

It is also noticed that the pressure shows symmetrical distribution; which suggest that problem has been simulated to satisfactory computational time to achieve a convergent solution. With different configurations analysed, adverse pressure gradient is larger in case of square cylinder with sharp corners and least for square cylinder with round corners as shown in Figure 8.

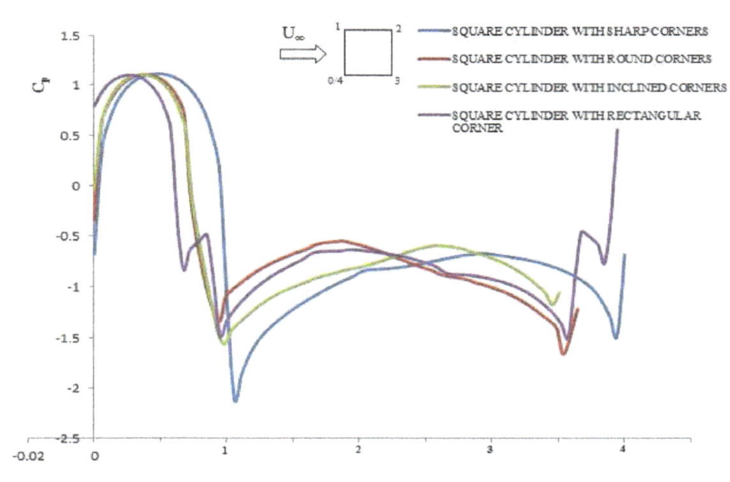

Figure 8: Coefficient of pressure variation along the cylinder surface

4.5 Pressure Variation in the near Wake region

In fluid flow analysis over bluff bodies, flow separates and creates eddies behind the object i.e. in the wake region. Eddies are alternatively shed from top and bottom of the body, inducing a transverse motion in the body. This periodic motion causes the body to vibrate and if the frequency of vibration matches with natural frequency of vibration of the body, resonance condition occurs and it may cause failure of the body or structure. Figure 9 presents the pressure distribution in the near wake region. Adverse pressure gradient is larger for square cylinder with sharp corners and least for square cylinder with round corners.

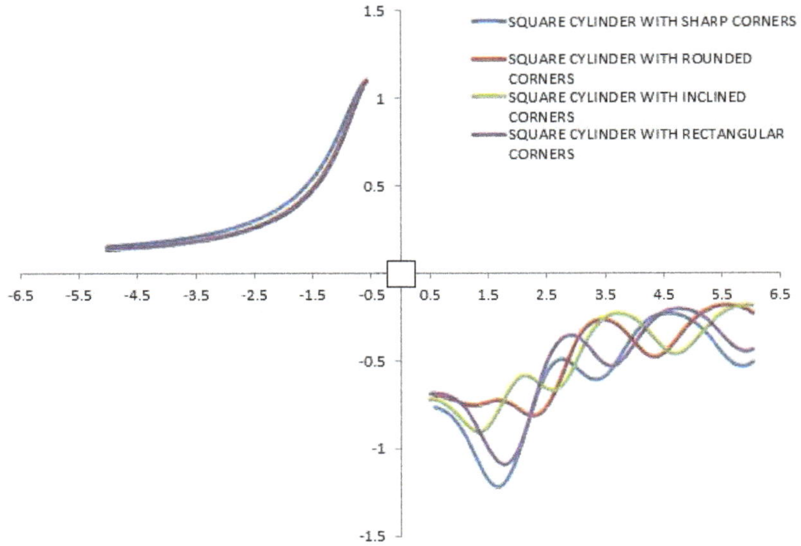

Figure 9: Variation of C_p in the near wake region

5. Conclusion

A numerical investigation of the flow around square cylinder with and without corner modifications at Reynolds number 150 has been presented. The simulation performed for flow past a square cylinder after grid refinement is in agreement with the theoretical results. The results of the numerical analysis around square cylinders with and without corner modification lead to the following conclusions:

- The tangential velocity of the Square cylinder is large when compared with corner Rounded and chamfered, and enlarges the separation area of square cylinder side face.
- Flow separation is delayed when the sharp corners of square cylinder are modified to round corners. Also flow reattaches before reaching the top right corner.
- When the fluid flows over the top surface of square cylinder, flow momentum is quite low due to viscous effects and is thus sensitive to changes of pressure gradient. Pressure gradient on this top surface is smaller for square cylinder with round corners and larger for square cylinder with sharp corners.
- Similarly pressure variation in the near wake region is larger for square cylinder with sharp corners and least for square cylinder with round corners.

Hence in the structures with Square cross-section, the sharp edge corners can be rounded-off so that magnitude of adverse pressure gradient can be reduced and hence flow separation can be delayed without use of any external source or structure.

References

[1] Sohankar, A, Norberg, C, and Davidson, L (1998): 'Low-Reynolds-number flow around a square cylinder at incidence: study of blockage, onset of vortex shedding and outlet boundary condition', International Journal of Numerical Methods in Fluids, 26(1), 39-56.00.

[2] INOUE, IWAKAM, and HATAKEYAMA (2006). "Aeolian tones radiated from flow past two square cylinders in a side-by-side arrangement." Physics of Fluids, 18(4), 046104.

[3] Ali, M.S.M., Doolan, C.J., Wheatley, V., (2009). Grid convergence study for a two dimensional simulation of flow around a square cylinder at a low Reynolds number. In: Seventh International Conference on CFD in The Minerals and Process Industries. CSIRO, Melbourne, Australia.

[4] Doolan, C.J (2009): 'Flat-plate interaction with the near wake of a square cylinder', AIAA Journal, 47,475-478.

[5] Bearman PW, Graham JMR, Obasaju ED, Drossopoulos GM (1984) The influence of corner radius on the forces experienced by cylindrical bluff bodies in oscillatory flow. Appl Ocean Res 6:83–89.

[6] H. Kawai (1998) Effect of corner modifications on aeroelastic instabilities of tall buildings, J.Wind Eng.Ind.Aerodyn. 74-76 719-729.

[7] Tamura T, Miyagi T, Kitagishi T (1998) Numerical prediction of unsteady pressures on a square cylinder with various corner shapes. J Wind Eng Indus Aerodyn 74:531–542.

[8] Tamura T, Miyagi T (1999) The effect of turbulence on aerodynamic forces on a square cylinder with various corner shapes. J Wind Eng Indus Aerodyn 83:135–145.

[9] Zheng W, Dalton C (1999) Numerical prediction of force on rectangular cylinders in oscillating viscous flow. J Fluids Struct13:225–249.

[10] R. Ajith Kumar, Chang Hyun Sohn, and B. H. Lakshmana Gowda (2009) Influence of corner radius on the near wake structure of a transversely oscillating square cylinder Journal of Mechanical Science and Technology 23. 2390~2416.

[11] J. C. Hu, Y. Zhou and C. Dalton(2006), Effects of the corner radius on the near wake of a square prism, Exp Fluid 40 106-118.

[12] Okajima, A., 1982, Strouhal numbers of rectangular cylinders. Journal of Fluid Mechanics 123, 379–398.

[13] Layukallo, T. and Y. Nakamura (2003), "Passive Separation Control on a Square Cylinder at Transonic Speed", Transactions of the Japan Society for Aeronautical and Space Sciences, 45(150), 236-242.

[14] Strykowski P I and Sreenivasan K R (2002): "On the formation and suppression of vortex 'shedding' at low Reynolds numbers", Journal of Fluid Mechanics 1990; 218: 71-107.

10

Fatigue Analysis of Drive Shaft

Sandeep Gujaran[1] and Shivaji Gholap[2]

[1] 1PG Scholar Department of Mechanical Engineering, SKN Sinhgad Institute of Technology & Science, Lonavala, Maharashtra India,

sandubms@gmail.com

[2] Assistant Prof. Department of Mechanical Engineering, Sinhgad Institute of Technology & Science, Lonavala, Maharashtra, India,

shivajigholap@yahoo.com

Abstract

The main objective of this analysis is to investigate the stresses& deflections of drive shaft subjected to combine bending & torsion. Then checking for fatigue life as well as comparing the results with analytical calculations to verify accuracy of the results. Drive shaft is a critical component used in paper converting machines. It carries a load of two vacuum rollers weighing around 1471N and rotates at 1000 rpm, also subjected to reaction force of knife cutter and gears. This shaft has key slots and at the area of change in cross sections giving rise to localize stress concentration. Hence there is a scope of analyzing this part to predict its fatigue life and damage.

Keywords: Fatigue Analysis, Shaft stress analysis, FEM analysis, shaft failure analysis

1. Introduction

A shaft is a rotating member usually of circular cross-section (solid or hollow), which is used to transmit power and rotational motion in machinery and mechanical equipment in various applications. Elements such as gears, pulleys (sheaves), flywheels, clutches, and sprockets are mounted on the shaft and are used to transmit power from the driving device (motor or engine) through a machine. In deciding on an approach to shaft sizing, it is necessary to realize that a stress analysis at a specific point on a shaft can be made using only the shaft geometry in the vicinity of that point. Thus the geometry of the entire shaft is not needed. In design it is usually possible to locate the critical areas, size these to meet the strength requirements, and then size the rest of the shaft to meet the requirements of the shaft-supported elements.

When you submit your paper print it in one-column format, including figures and tables. In addition, designate one author as the "corresponding author". This is the author to whom proofs of the paper will be sent. Proofs are sent to the corresponding author only.

The deflection and slope analyses cannot be made until the geometry of the entire shaft has been defined. Thus deflection is a function of the geometry everywhere, whereas the stress at a section of interest is a function of local geometry. For this reason, shaft design allows a consideration of stress first. Then, after tentative values for the shaft dimensions have been established, the determination of the deflections and slopes can be made.
Most shafts are subjected to fluctuating loads of combined bending and torsion with various degrees of stress

concentration. For such shafts the problem is fundamentally fatigue loading. Failures of such components and structures have engaged scientists and engineers extensively in an attempt to find their main causes and thereby offer methods to prevent such failures

1.1 Shaft stresses

Bending, torsion, and axial stresses may be present in both midrange and alternating components. Axial loads are usually comparatively very small at critical locations where bending and torsion dominate, for a rotating shaft with constant bending and torsion, the bending stresses completely reversed and the torsion is steady. The fluctuating stresses due to bending and torsion are given by

$$\sigma_Z = \frac{M}{Z} = \frac{32M}{\pi\, d^3} \quad (1)$$

The stress analysis process for fatigue is highly dependent on stress concentrations. Stress concentrations for shoulders and keyways are dependent on size specifications that are not known the first time through the process. Fortunately, since these elements are usually of standard proportions, it is possible to estimate the stress concentration factors for initial design of the shaft.

1.2 Deflection Considerations

Deflection analysis at even a single point of interest requires complete geometry information for the entire shaft. For this reason, it is desirable to design the dimensions at critical locations to handle the stresses, and fill in reasonable estimates for all other dimensions, before performing a deflection analysis. Deflection of the shaft, both linear and angular, should be checked at gears and bearings. Allowable deflections will depend on many factors, and bearing and gear catalogs should be used for guidance on allowable misalignment for specific bearings and gears.

For shafts, where the deflections may be sought at a number of different points, integration using either singularity functions or numerical integration is practical. In a stepped shaft, the cross sectional properties change along the shaft at each step, increasing the complexity of integration, since both M and I vary. Fortunately, only the gross geometric dimensions need to be included, as the local factors such as fillets, grooves, and keyways do not have much impact on deflection.

A deflection analysis is straightforward, but it is lengthy and tedious to carry out manually, particularly for multiple points of interest. Consequently, practically all shaft deflection analysis will be evaluated with the assistance of software. Any general-purpose finite-element software can readily handle a shaft problem. This is practical if the designer is already familiar with using the software and with how to properly model the shaft. Special-purpose software solutions for 3-D shaft analysis are available, but somewhat expensive if only used occasionally.

For hand calculation, shaft deflection can be calculated by using double integration formula,

$$\frac{d^2y}{dx^2} = \frac{M}{EI}\mathrm{dx} \quad (2)$$

1.3 Overview of Fatigue

The majority of component designs involve parts subjected to fluctuating or cyclic loads. Such loading induces fluctuating or cyclic stresses that often result in failure by fatigue. About 95% of all structural failures occur through a fatigue mechanism. The damage done during the fatigue process is cumulative and generally unrecoverable, due to the following:

It is nearly impossible to detect any progressive changes in material behavior during the fatigue process, so failures often occur without warning.
Periods of rest, with the fatigue stress removed, do not lead to any measurable healing or recovery.
Fatigue, or metal fatigue, is the failure of a component as a result of cyclic stress. The failure occurs in three phases: crack initiation, crack propagation, and catastrophic overload failure. The duration of each of these three phases depends on many factors including fundamental raw material characteristics, magnitude and orientation of applied stresses, processing history, etc. Fatigue failures often result from applied stress levels significantly below those necessary to cause static failure.

1.4 Finite Element Method (FEM)

FEA as applied in engineering is a computational tool for performing engineering. It includes the use of mesh generation techniques for dividing a complex problem into small elements, as well as the use of software program coded with FEM algorithm. In applying FEA, the complex problem is usually a physical system with the underlying physics such as the Euler-Bernoulli beam equation, the heat equation, or the Navier-Stokes equations expressed in either PDE or integral equations, while the divided small elements of the complex problem represent different areas in the physical system.
It includes the use of mesh generation techniques for dividing a complex problem into small elements, as well as the use of software program coded with FEM algorithm. In applying FEA, the complex problem is usually a physical system with the underlying physics such as the Euler-Bernoulli beam equation, the heat equation, or the Navier-Stokes equations expressed in either PDE or integral equations, while the divided small elements of the complex problem represent different areas in the physical system.

---

2. DESIGN OF DRIVE SHAFT

Figure 1 Loaded drive shaft assembly

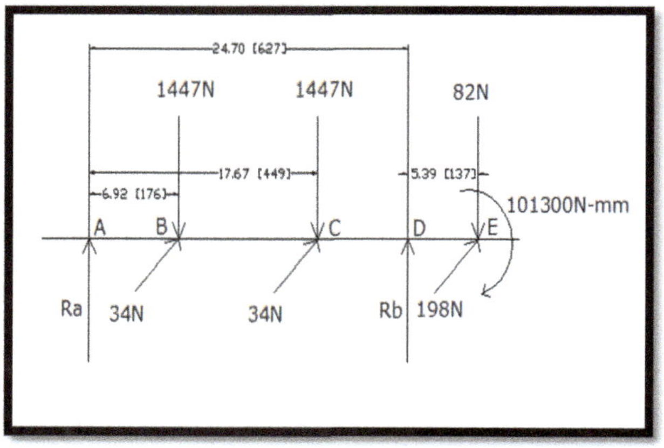

Figure 2 free body diagram

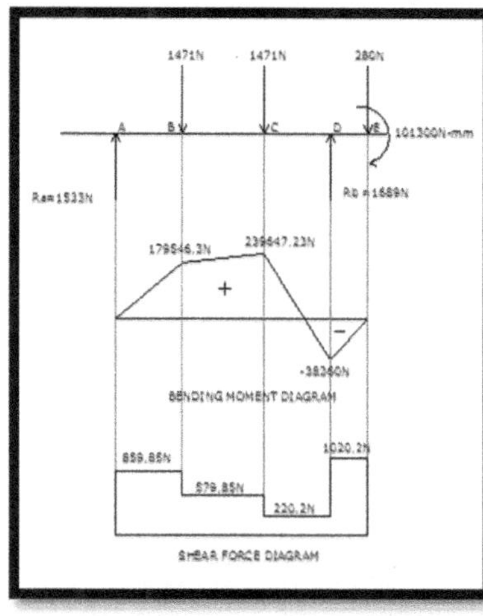

Figure 3 Bending & shear force diagram

Table 1: Shaft material properties

Material	Yield strength in Mpa	UTS in Mpa	Hardness HB	Shear modulus in Gpa	Young's modulus in Gpa
1040 Medium carbon steel	370	585	185	80	200

2.2 ANALYTICAL CALCULATIONS

2.2.1Min. diameter of shaft under combined bending and torsion

$$D = \sqrt[3]{\frac{5.1}{P_t}\sqrt{(K_m M)^2 + (K_t T)^2}} \qquad (3)$$

d =30 mm, taking factor of safety = 2, d = 60mm

2.2.2 Deflection of shaft

Bending moment at any section distance x from A is given by

$$\frac{d^2y}{dx^2} = \frac{M}{EI}dx \qquad (4)$$

By double integration and applying boundary conditions we get

@ B, y = 0.23mm, @ C, y = 0.247mm and @ E y = 0.02mm

Deflection is too great at location C, the needed shaft diameter is given by equation:

$$d_{new} = d_{old} \left| \frac{n_d \, y_{old}}{y_{all}} \right|^{1/4} \quad (5)$$

Where n_d is the design factor = 0.5

y_{all} is the allowable deflection at that location = 0.0523 mm

d_{new} = 73 mm

So we will increase the diameter at the point where vacuum rollers are mounted to 73mm.

2.2.3 Slop of shaft

$$\frac{dy}{dx} = \frac{M}{EI}dx = \Theta \quad (6)$$

By integration and applying boundary conditions we get

@ B, Θ = 0 .00087rad, @ C, Θ = 0.00013 rad and @ E Θ = 0.00102 rad

For tapered roller bearing max allowed slop is 0.0005 to 0.0012 and calculated slops are less than this, hence design is safe.

2.2.4 Bending and shear stress acting on shaft

$$\sigma = \frac{32M_b}{\pi d^3}, \tau = \frac{16M_t}{\pi d^3} \quad (7)$$

2.2.5 Equivalent stress/von Mises stress

$$\sigma_{eq} = \sqrt{\sigma_{bmax}{}^2 + 3(\tau_{\max})^2} \quad (8)$$

σ_{eq}= 33.4 Mpa

2.2.6 Endurance strength

$$S_e = K_{load} \times K_{size} \times K_{surf} \times K_{temp} \times K_{reliability} \times K_d \times K_{stress\ concentration} \times S_e' \quad (9)$$

K_{load} = 1 (Bending Load),

$K_{size} = 1.24d^{-0.107}$ for $2.79 \leq d \geq 51$ mm = 0.78

$K_{surf} = 4.51 S_{ut}{}^{-0.265}$ = 0.82

$K_{temp} = 1$ for $T \leq 450°C$

$K_{reliability} = 0.90$

$K_d = \frac{1}{Kf} = \frac{1}{1.89} = 0.53$

$K_{stress\ concentration} = 1+q\ (k_f-1) = 1+0.82(1.89-1) = 1.73$

$S_e = 167$ Mpa

2.2.7 Number of Cycles

Estimated Fatigue Life

Infinite Life- 1000RPM (Average operating speed)=16.66 cycles/second
5 year life @ 7 hour operating time per day approximately 48384000 seconds of use
16.66 x 48384000=8.06 x 10^8 cycles to failure for infinite life

$$N = \left(\frac{Sf}{a}\right)^{\frac{-1}{b}} \quad (10)$$

$$a = \frac{(fS_{ut})^2}{S_e} \qquad b = -\frac{1}{3}\log\left(\frac{fS_{ut}}{S_e}\right) \quad (11)$$

2.2.8 Fatigue Damage Factor

Fatigue damage factor $= \frac{\text{Expectd life cycle}}{\text{Actual life cycle}} = \frac{8.06 \times 10^8}{1.21 \times 10^{10}} = =0.4 <1$

3. FEM study using Solidworks Simulation

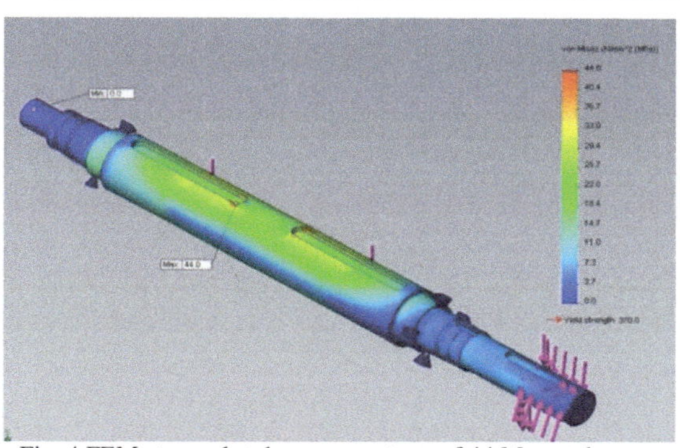

Fig. 4 FEM stress plot shows max stress of 44 Mpa at the center

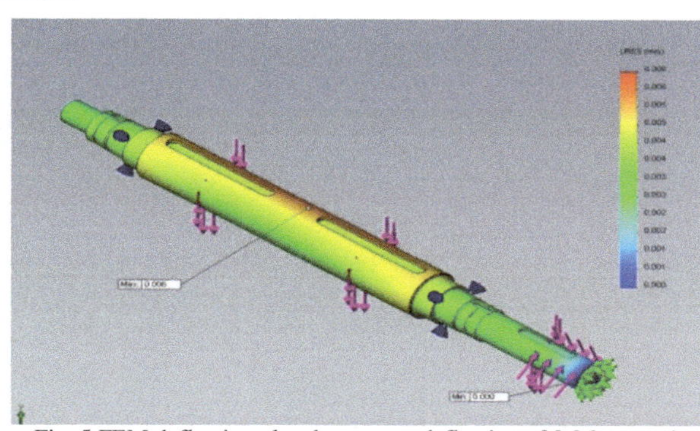

Fig. 5 FEM deflection plot shows max deflection of 0.06mm at the center

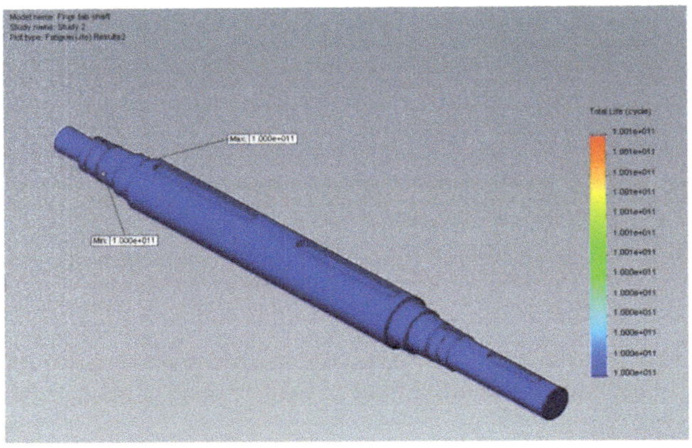

Fig. 6 Fatigue Life plot shows max. Fatigue life of 1x10^11 cycles

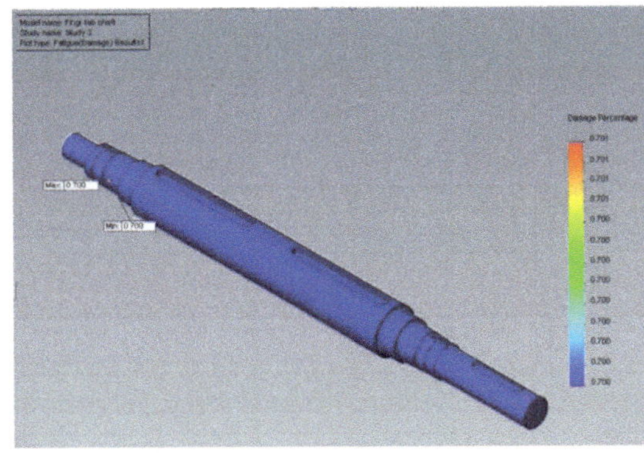

Fig.7 Fatigue damage plot shows 0.7 which is less than unity

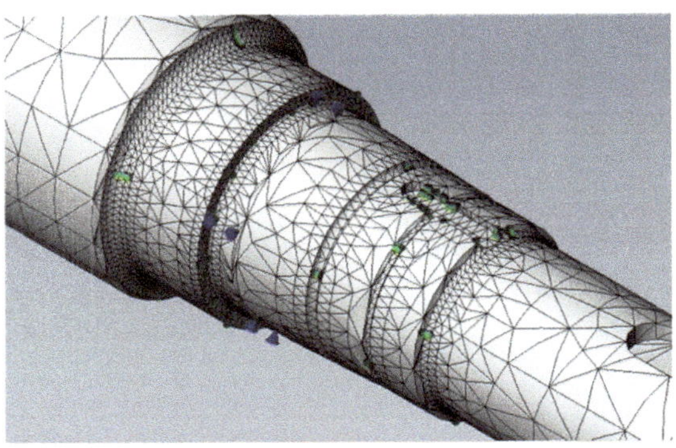

Fig. 8 Close look at the mesh control at the critical junctions analysis

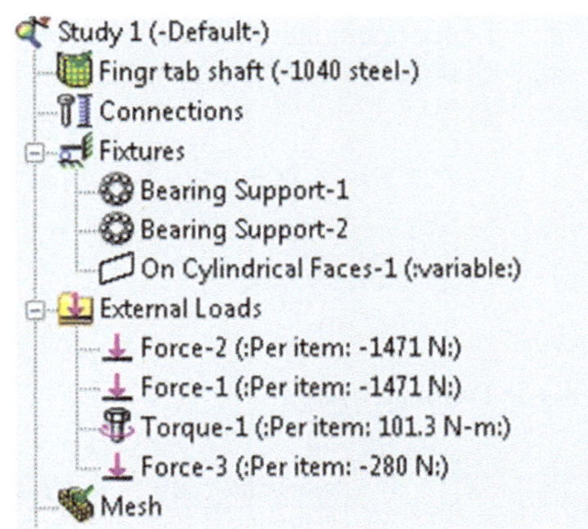

Fig. 9 Feature tree showing desired inputs for FEM

4. CONCLUSIONS

Table 2: Comparison of results

Parameter	Analytical results	FEM results
Max Stress in Mpa	33.41	42
Deflection in mm	0.029	0.06
Fatigue Life in cycles	8.0 x 10^8	1 x 10^11
Fatigue Damage	0.4	0.7

Based on FEM results it is found that max stresses found at the bearing junction because of change of cross section and small radius at the junction.

Geometry of shaft is modified by introducing a step at both side of the shaft before bearing junction and giving max. Radius, so that stress concentration could be minimized at the bearing junction.

After rechecking stress levels reduced by major amount and fatigue life and damage is calculated by Simulation and they found to be very much within the limits.

5. References

[1] Joseph E Shigley Mechanical Engineering Design, Sixth Edition.
[2] Industrial press inc, Machinery Handbook, Newyork, 27th edition
[3] Depan Marudachalam, Optimization of shaft design under fatigue loading, IJSER, Volume 2, Issue 8, August-2011
[4] R.A. Gujar, Shaft design under fatigue loading, IJERA, Volume 3, Issue 4, Jul-Aug 2013.
[5] D.K. Padhal, Analysis and failure improvement of shaft of gear motor, RIIJOAS, Vol 3, Issue 4, July 2013
[6] E. Rusinski P. Harnatkiewicz, Caterpillar drive shaft damage analysis, Mechanical Engineering, Vol VIII, 2008, No3.
[7] Pushpendra Kumar, Fatigue Analysis of connecting rod using FEA, IJARSE, 2012, Vol. No.1, September.
[8] Calalettin, Fracture and Fatigue analysis of an Agitator Shaft, KRAAGAC, July, 2002, IZMIR.

"Structural Analysis and Design of Bag Palletizer Gantry"

Dhananjay Patel

Department Of Mechanical Engineering, Vishwakarma Institute Of Technology, Pune 411 037, Maharashtra, India.
Email: dhananjaykpatel17@gmail.com

ABSTRACT

Modern technology period can't be imagined without various material handling equipments. Gantry cranes are amongst one of the material handling equipment which finds wide application in different fields of engineering. Due to incredible all around economical development, the operation rate of gantry cranes has increased year by year. Many gantry cranes are often use beyond their capacity in India. So Analysis of the main structure is very important and essential. The purpose of this study is to analyze the stress and displacement condition in the structure of Bag palletizer gantry crane, presenting a fast and evaluated computer aided solving method for complex static structures. The analysis of the stress and displacement of the structure of gantry crane is made using the Hypermesh Radioss. The research performed allows the evaluation of the stress state, pointing out the critical areas and measures which are imposed in order to increase the toughness of the power structure for the gantry crane .Thus , the material use could be reduce without more than the limits of the most suitable conflict.

Keywords: Gantry, crane, Palletizer, HypermeshRadioss, conflict.

INTRODUCTION

Material handling involves "short-distance movement that usually takes place within the confines of a building such as a plant or a warehouse and between a building and a transportation agency." It can be used to create "time and place utility" through the handling, storage, and control of material, as distinct from manufacturing.

Bag palletizer is communally known as gantry robot. Which is an material handling equipment. A bag palletizer consist of a manipulator mounted onto overhead system that allows moment across horizontal and vertical plane. These gantry robot is use to pick a bag form conveyer and put it on wooden pallet in sequential manner.

Bag Palletizer Gantry(BPG) crane are amongst one of the material handling equipment which find wide application in different fields of engineering. Gantry cranes are industrial machine that are mainly used for material movements in construction sites, production halls, assembly lines, Storage areas and similar places. Their design features vary broadly according to their major operational specification such as type of motion of crane structure, weight and type of the lode, location of the crane, geometric feature, operating regime and environmental conditions. Consideration of available technology that is mainly based on the accumulated previous experience is important for better performance, higher safety and more reliable designs. Now a day's cold formed member are use as primary load bearing structure. Main reason behind the extensive use of cold formed member is that these are easy and cheap to fabricate, need minimal maintenance due to zinc coating.

In the recent decade the fields of cold formed thin walled steel structures have been among the busiest research area at the department of steel engineering. The design method and practical approaches and formulae that are based on previous design experience. It is believed that computer automated access to these standards with preloaded interpretation and guidance rules increase speed and reliability of the design procedure and increase efficiency of the gantry crane designers.

Bag Palletizer Gantry equipment has been traditionally design using some standards with factor of safety included into the design. This can lead to over design of component. This project having concept model of bag palletizing gantry crane in CATIA. The Structural analysis carried out in Hypermesh Radioss solver for a 100 kg load condition Bag palletizer gantry.

1.1 Problem Statement

Bag Palletizer Gantry is specially used to pick and place type material handling equipment. It is an modern industrial machine use in Bag palletization process. The working range varies from few kilograms to many tons. In our case for Bag palletization of 100 kg of bag is pick form conveyer and put it on wooden pallet. The available gantry system are made up of extruded aluminum profile. This aluminum profile gantry has an problem of high cost in Indian market. Also due to aluminum profile the weight of gantry is also high and which affect on working speed of machine. So our aim is to introduce light weight, cost efficient product in market. At the same time due to light weight machine increase working speed of machine. Especially,3D Modeling of bag palletizer gantry and investigating deflection and stress distribution on gantry parts are critical for effective working of new designed Bag Palletizer Gantry.

1.2 Objective and Scope

The main objective of this thesis study is to investigate stress and deformation of bag palletizer gantry parts of Cold rolled mild steel material, induced by load and boundary condition.

The elements of the main objective are:

1) To determine new material and cross section for design of Bag Palletizer Gantry.
2) To create three dimensional solid model of Bag Palletizer Gantry machine in CAD software.
3) To perform static loading simulation of the three dimensional model using FEA software, with the selection of suitable material for Bag Palletizer Gantry at given load and boundary condition.
4) To investigate and analyze mechanical effect of load on Bag Palletizer Gantry components stress and deformation.

The scope of this thesis is to investigation of required data for design of Bag Palletizer Gantry component. Find out stress and deformation of Bag Palletizer Gantry solid model with the help of static analysis in FEA environment.

1.3 Outline of Thesis

General information about the thesis is provide in the first chapter. Following the introductory chapter, a comprehensive literature review is present in the second chapter. Literature review basically contain previous studies about gantry, various cold rolled mild steel section and finite element analysis.

The third chapter discusses the basic of Bag Palletizer Gantry preliminary information, with material selection study. The chapter four having study of methods and approaches , study of gantry cost model and approach toward concept is discuss in quality function development, followed by Modeling software and finite element analysis study.

The next chapter five contain gantry requirement and basic data required to design and house of quality design for bag palletizer. Followed by chapter six having basic design procedure toward part design and selection of parts. Actual model design is carried out in chapter seven with help of reference model and finally finite element analysis is done Hypermesh Radioss environment in chapter eight. Numerical and analytical result calculate in graphical visual of simulation result and discussion is subject of ninth chapter.The thesis study end with conclusion and recommendation part in tenth chapter. Reference is the last section of this thesis work.

2. LITERATURE REVIEW

Gantry are particularly suited to lifting very heavy objects. Huge gantry have been use for steep building, where the object like ship engine to lifted and move over the ship. The first over gantry built in 1969 are **Samson and Golith** in Belfast,Northern Ireland. This gantry having span 140meter and can lift load up to 840 tones. The same type of gantry are use in ware houses to loading and unloading of material .[3]

Fig2.1.1. Samson and Golith in Belfast,Northern Ireland [3]

Gantry crane are basically made up of aluminum and mild steel structure.[2] But small size of Bag Palletizer Gantry are made up of aluminum profile [6].The main disadvantage of aluminum bag palletizer gantry is their high cost. Because, extruded aluminum profile is expensive in manufacturing and machining also. **Bernarda and Ferreira[2]** introduce a method to evaluate and develop new gantry concept. The evaluation is done with the help of House of quality method. Which is best concept design method in early industrial development with the help of this method Bag Palletizer Gantry concept design time and all design aspect. Every possibility of product development is provided with House of quality study. In 1979 Toyota motors reduce 20% of starting cost and in 1984 it turns to 61% using House of quality study. This method for gantry design and development gives an engineering focus toward concept design.

Cold Rolled C-section of mild steel member in gantry design is considered because of **Gabor and Jakobs [1]** research work. An experimental program on compression member made up of mild steel cold rolled section. Various C-section analyses which were not studied previously. Mild steel section are capable various load condition and structural strength capacity is analyze in this research work. Gantry structure analyze by Finite element analysis and changing various material property to improve solidity of member. An numerical and analytical results varies proportionally while changing steel property in main longitudinal girder part in gantry model. Distortional buckling of C section is tested by **Cheng yu and Ben Schafer[4]** in The American iron and steel institute, The experimental validation of wide verity of mild steel C-section. This study indicates that distortion buckling is most likely failure mode. This study contain test setup for C-section and various C-member are tested for buckling and all those condition are verified with finite element analysis with same lode and constrain use in test setup. The result of the study shows reasonable and conservative prediction for both local and distortional buckling strength of beam .Which is use for member selection for gantry design for particular load condition.

C.Alkin,C.E.Imrak,H.Kocabs [5] work on solid modeling and finite element analysis of gantry crane. While analysis material property of mild steel is used and evaluate stress and deflection. The analysis study concludes that four node quadratic shell element is gives accurate result than four node tetrahedral element use in mashing model in FEM software. This study help while calculating concept model with its applied load and boundary condition **J.I.Lergea[10]** studied in modeling and simulation research on metal structure on bridge crane. The main objective of this study is to reduce structural weight of girder of gantry crane. For these purpose several model of girder structure have been designed. The new lighter model is design is analyzed with load condition. The structural material used in this model is economical too. This designed model have been lose its weight by holes in structure. That can reduce structural weight without affecting structural strength of gantry structure.

3. BASIC OF BAG PALLETIZER GANTRY

Bag palletizer is communally known as gantry robot. Which is an material handling equipment. A bag palletizer consists of a manipulator mounted onto overhead system that allows moment across horizontal and vertical plane. These gantry robot is use to pick an bag form conveyer and put it on wooden pallet in sequential manner.

The Bag Palletizer Gantry contains manipulator unit which is mounted on Girder span, on which an manipulator assembly moves linearly. The Girder section is attached with Carriage plate at both the ends of Girder. The end carriage plate is attached to Runway beam one at each end. On which Girder assembly moves linearly. Runway beam is simply supported by two support column at each end of runway beam. In this manner Bag Palletizer Gantry system structure is to be assembled.

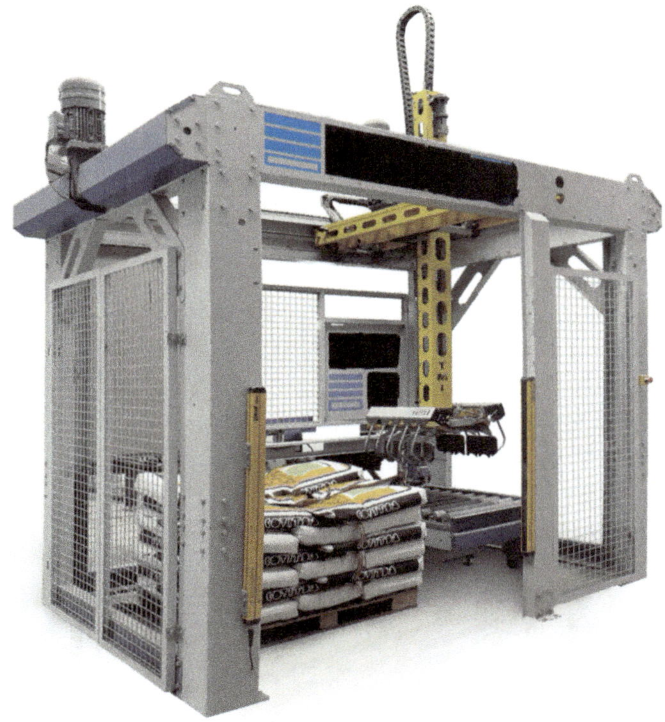

Fig.3.1.1 Bag Plalletizer Gantry of company TMI Bagging and Palletizing [13]

The movement of material from the place where it is to the place where it is needed can be time consuming, expensive, and troublesome. The material can be damaged or lost in transit. It is important, therefore, that it be done smoothly, directly, with the proper equipment and so that it is under control at all times. The several factors that must be known when a material handling system is designed include:

1. Form of material at point of origin, e.g., liquid, granular, sheets,etc.
2. Characteristics of the material, e.g., fragile, radioactive, oily,etc.
3. Original position of the material, e.g., under the earth, in cartons,etc.
4. Flow demands, e.g., amount needed, continuous or intermittent,timing, etc.
5. Final position, where material is needed, e.g., distance, elevationdifferences, etc.
6. Handling equipment available, e.g., devices, prices, reliability,maintenance needs, etc.

Other factors to be considered include:

1. Labor skills available
2. Degree of mechanization desired
3. Capital available
4. Return on investment
5. Expected life of installation

Since material handling adds expense, but not value, it should be reduced as much as possible with respect to time, distance, frequency, and overall cost. A straight steady flow of material is usually most efficient. The use of mechanical equipment rather than humans is usually, but not always, desirable depending upon the duration of the job, frequency of trips, load factors, and characteristics of the material. When equipment is used, maximizing its utilization, using the correct equipment, proper maintenance, and safety are important considerations. The proper material handling gantry equipment can be selected by analyzing the material, the route it must take from point of origin to destination. The importance of gantry is that it helps productivity and increase profitability of an industry. Many industries go out of business because of insufficient material handling gantry. In many instance it is seen that competing industries are using same or similar equipment,and one who uses improved gantry system stay ahead of their competitors.

A well designed gantry system attempts to achieve the following:

- Improve efficiency of production system by ensuring right quantity of material delivered at the right place at the right time most economically.
- Cut down indirect labour cost.
- Reduce damage of material during storage and movement.
- Maximize space utilization by proper storage of material.
- Minimize accident during material handling.
- Reduce overall cost by improving design of Bag Palletizer Gantry.

All the above points clearly show the importance of gantry system in material handling in an industry.

4. METHODS AND APPROACHES

Firstly, a Cost Model applied to Gantry structures will be discussed. Then the applied approaches as Quality Function Deployment and House of Quality, and lastly the CAD and FEA software tools will be introduced.[2]

4.1 Gantry Cost Model

The main goal of this work is to develop an mild steel Bag Palletizer Gantry system, which has lower cost than the present aluminum gantry. Since the cost is a key factor in terms of competiveness for the final product, it is very relevant to be able to access the costs thorough each step during the product fabrication.

The development of a cost model has two big goals. First, by applying it to the current solution, the cost distribution in each step might be visible and it can be evaluated where there is potential for improvement during the conception of the new Gantry solutions. The second goal is to predict the cost by comparison of different solutions during the concept phase of this work. This tool can have a very important role during the design stage, since is through decisions taken in the primary product development phases (as concept proposals and the evaluation between them) that the highest amount of the final product cost as well as quality are determined.

The cost model used in this work divides the Gantry cost into five major categories: material (cold rolled mild steel), parts (connection plates, screws and nuts), assembly, Power drive (motor, bearing ,HTD belt) and production (operations after cold rolling of mild steel sheet ; cutting and drilling).[2]

4.2 Quality Function Deployment (House of Quality)[11]

This section will address the concept of Quality Function Deployment (QFD) and its importance in the early stages in product development.

QFD was developed in Japan in 1966, by Yoji Akao, a planning specialist also developer of the Hoshin Planning. QFD is an approach used in order to achieve a product that meets the exact needs of the customer. It is a tool (matrix) used to convert subjective wants and needs into objective engineering specifications. Thus, the major goals are to rate and prioritize the customers' demands, convert them into technical specifications, and lastly be able to deliver a quality product aiming for the ultimate customers' satisfaction.

There are four main phases, each one corresponding to a matrix, considered in this approach Product Planning, Product Design, Process Planning and Process Control. The first phase, also called House of Quality, consists of the correlation between the customer's needs or requirements with the technical descriptors, which the company will use to meet those demands. The second phase, Product Design, is used for the concept generation and selection of those which meet the customers' requirements the best. The third phase, Process Planning, as the name suggests corresponds to the planning of the manufacturing operations. Lastly, the Process Control consists of the control procedures done during the production to assure the product's target quality levels.

The application of the QFD in this work will only focus on the first phase, the House of Quality. During the evaluation of different concepts it will be important to have a rating of the performance parameters, which are governed by the requirements and needs. The House of Quality, by correlating the customer's needs and wants (the requirements) with technical characteristics, makes possible to have a relative weight for all those parameters used to compare concepts.

The House of Quality is a matrix that can be summarized in six major sections. A typical House of Quality has the layout presented in the figure below, with the different numerated sections.

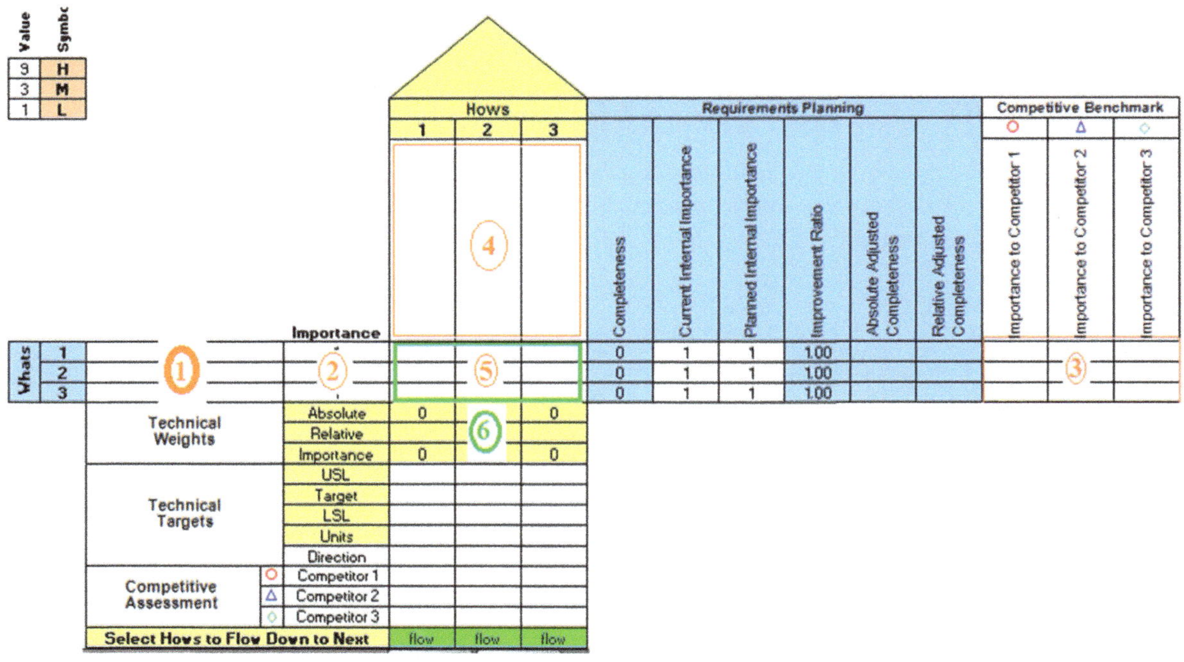

Fig.4.2.1 Outlook of House of Quality

Below the contents of the different parts will be summarized.

1) Customer Requirements: This section is the list of the customer's needs and wants that correspond to the project's requirements.

2) Importance Ratings: The customer shall rate each requirement in a scale from 1 to 5 according to the importance.

3) Customer Rating (of the company and competition): In this section, the customer can rate the degree of satisfaction (1 to 5) of each need, perceived from the company or/and from the competition.

4) Technical Descriptors: This section consists in the attributes that will quantify and measure the customer's requirements. In other words, these are the means to achieve customer satisfaction.

5) Relationship Matrix: In this section is quantified the strength of the relationship between the requirements and the technical attributes. Therefore, it measures the influence of each attribute in the fulfillment of each requirement. The relation can the Strong, Medium or Weak, and be rated with 9, 4 or 1, respectively.

6) Absolute and Relative Importance: This is the section that reveals the most important technical attributes that most matter for the customer. The Absolute Importance assessment is the product of the value in each cell of the Relationship Matrix (6) with the correspondent Importance Rating (2).

In the last section of the House of Quality, the Relative Importance for each attribute of the design is calculated. This will be significant to evaluate different concepts in a qualitative way, since the different technical parameters don't have the same importance to achieve maximal customer satisfaction.

4.3 CAD Software

This section addresses the importance of CAD software tools and its function within this work.CAD means Computer Aided Design. These software tools allow the modeling of technical designs. They are used broadly in all the engineering departments but also on many other subjects. CAD software permits to draw 2D or 3D models. By modeling the product to be developed one can preview and see the real dimensions and the couplings between the different parts of the total assembly and some problems can be identified already in the modeling stage. It consists in a very good tool to see a virtual version of the real product before it actually exists, that helps the designer but can also be used for presenting the product to the public. This software usually allow dimensional changes after the modeling being complete with the update of the total structure, which is very convenient during the optimization of the model.

The CAD models, besides being important for the preview of the assembly between all the structure's parts and giving a visual image of the product to be produced or developed, they can be used in the subsequent stages of analysis. Normally, the Finite Element Analysis (FEA) software's allow the import of the CAD files with the model to be analyzed. CAD models can be used in FEA to predict the structural behavior.

CAD files will be vastly used during this work. It will be used to model the different parts and to build up the Gantry's assembly and with this compare different alternatives. Finally they are used to make a structural analysis, which is required to access the performance of the solutions.

4.4 FEA Software

The FEA software will be the tool used to perform the structural analysis and predict the behavior of the structure under the work loads.

The Finite Element Method decomposes the structure into small elements that are connected at nodes, through the application of a mesh. The mesh will transform a continuous domain into a finite domain that enables the calculation of simultaneous algebraic equations. This method requires the application of boundary conditions such as fixed or pinned constrains and the application of external loads and/or gravity in order to consider the self-weight of the structure.

The FEA can be performed in a structural application but also in thermal or fluid flow problems. As this work is focused on the design of a Gantry system, only structural analysis will be considered.

The structural analysis can be linear or non-linear. If the interest of the analysis is to preview the behavior of the structure in the elastic domain then is used a linear analysis. On the other hand, if the work loads are probable to overcome the yield stresses in some areas of the model, some hardening will happen and the material properties will differ along the model, therefore a non-linear analysis shall be performed. In this work, the structure is expected to work below the plasticity domain, or in other words, work only in the elastic domain, so the analyses to be performed will be linear. The goal of these linear analyses is to be able to preview the main results such as the maximum displacements and maximum stress.

5. BAG PALLETIZER REQUIREMENT

The product development phase has to be conducted aiming to meet the requirements that the product has to fulfill. This way, through the restrictions to the design it is possible to achieve a solution close to the optimum one. Without requirements, it would not possible to define a certain number of solution possibilities and the design options would be infinite.

The design of a product can have three types of implemented features or characteristics: the *Must Have*'s, the *Should Have*'s and the *Could Have*'s. A *Must Have* corresponds to a requirement, which defines the minimum level of customer's satisfaction. The *Should Have*'s correspond to additional features that cover extra needs of the client and add value to the product. The implementation of *Should Have*'s requires a proportional effort and resource consumption to the expected customer's satisfaction and corresponds to the expected quality of the product. Lastly, the *Could Have*'s are extra features that are added to the product to differentiate it from the competition and make the customer recognize its added value. The implementation a *Could Have* requires an exponential effort comparing to the customer's satisfaction and work a delighter.

The graph presented below (*Kano Model*) distinguishes qualitatively the three types of implementations that can be done during the design of a product. It relates the degree of implementation to the customer's satisfaction.

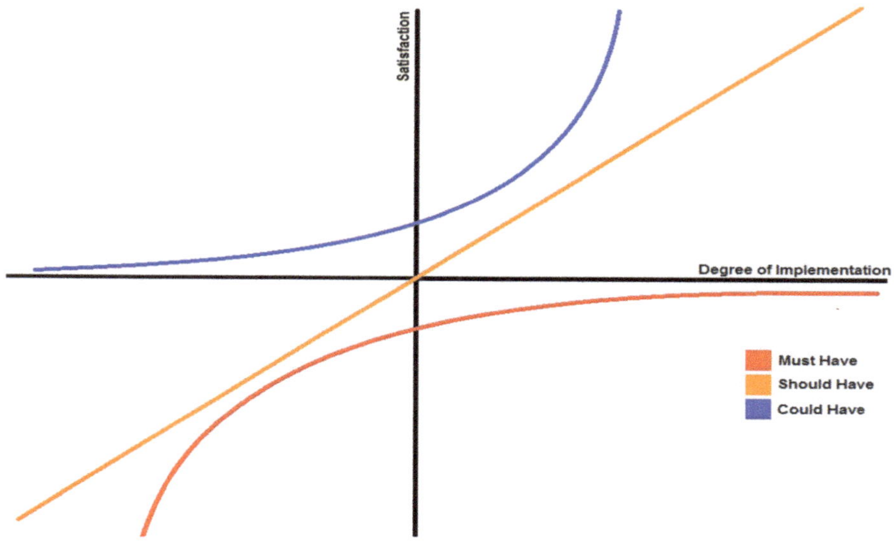

Fig.5.1.1 Graph of Kano model

The next section will cover the requirement that the Bag paletizer Gantry's design must meet. These requirement mostly stated by Company and customers. Therefore, the External Requirements, given by the customer that is obliged to comply with the norm, and the Internal Requirements, given by company, where are addressed issues that don't concern directly the customer, will be distinguished.
Each requirement will be categorized as a *Must Have* (M), *Should Have* (S) or *Could Have* (C).

5.1 External Requirement

This section will deal with the requirement stated by the customer. It is relevant to refer the most of the external requirement are beyond the customer needs .These requirements are help to improve importance of the Bag Palletizer Gantry in customer point of view.

Below, the requirements given by the customer are listed.

A Gantry is design must be material efficient aiming to light-weight structure.

Girder length : 2m to 2.5m

Height of support column : 1.5m to 2m

Runway beam length : 2m to 2.5m

Size of Pallet: 1200×1200×250mm

Bag size : 400×600×150mm

Bag layers : 6

Conveyer width:400mm

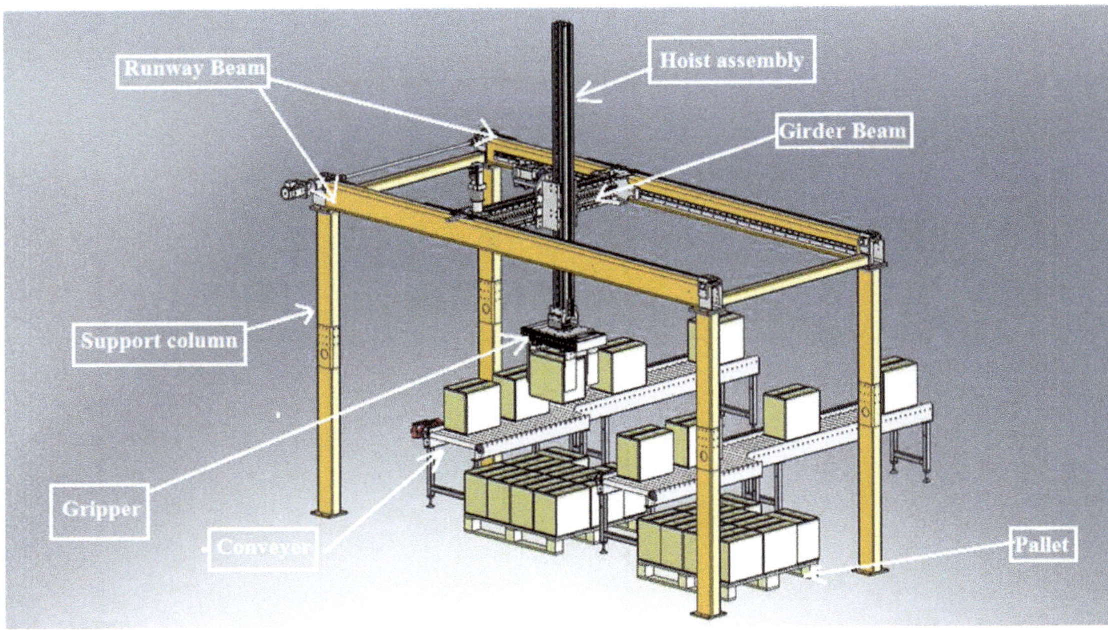

Fig.5.1.2 Gantry's Girder, Runway, Support column & Gripper CAD Model.[6]

Loads:

- Gantry self-weight

-Weight of Bag

Connection between Structural elements:

-Nut & bolt ,arc welding .

Environmental Conditions:

Gantry must be satisfy any environment like Powder plant, food industry ,Cement industry.

5.2 Internal Requirement

This section will address the requirement that company dictates.

Parts manufacturing process: Cold rolled sheet metal

Minimize the production cost:

The design of C-section should consider cold rolling process in order to provide better surface finish and property. The design should minimize the number of operations after cold rolled C-section manufacturing, means cutting ,drilling etc.

Minimize the assembly cost:

The design should consider profiles that simplify the assembly and also minimize the total number of connections, in order to keep down the assembly cost.

Minimize the Drive system cost:

The Drive system use in gantry system must be selected on basis of cost. In compression with rack and pinion drive, the belt and pulley drive is cost efficient system and high speed working is achievable[12].

5.3 House Of Quality

A product must comply with the requirements, mainly the customer's needs and demands, in order to be valuable and desired. The translation of the requirements into measurable parameters that define the product is a very important task. The requirements are achieved by the performance of the technical characteristics of the product. This way, it is very important, firstly, to decompose a Bag Palletizer Gantry into its several technical parameters. After that, list all the external and internal requirements and rate them accordingly with their importance (from 1 to 5). Then, it is needed to relate each requirement with each technical parameter (technical descriptor) in one of the three relations: Strong, Medium or Week. This way, through the House of Quality matrix, introduced in the fourth chapter, it can be known the relative importance or weight of each technical parameter regarding the ability of fulfilling all the requirements.

Since the requirements were introduced in the previous sections, below the technical parameters that allow characterizing a Gantry in terms of its structural configuration are listed.

-Number of Beam

-Number of Column

-Number of Connections (Nut Bolts)

-Number of Belt and pulley, Motors

The figure below presents the House of Quality matrix with the requirements' importance rated, as well as their relationship with each technical descriptor. These two values will originate the absolute and consequently the relative importance of each technical parameter, which is the goal of this exercise. The House of Quality matrix is a more complex tool, which represents just a step in the Quality Function Deployment approach. Therefore, it also represents the impact of the implementation of each technical descriptor in each other, regarding the correspondent directions of improvement. These relations will define the organizational difficulty in implement each technical descriptor. The matrix also has a field where the customer can evaluate currently the perceived capacity of the company and also the competitors to fulfill each requirement (this evaluation was assumed, because there was no contact to specific customers). Regarding the metrics, it was not considered any baseline values, because it is not being studied a specific Bag Palletizer Gantry size. Therefore, in short, the relevance of this implementation is to get the relative weights of each technical parameter.

Value	Symbol
9	H
3	M
1	L

Whats		Importance	How 1: Beam	How 2: connections	How 3: Bolts & Nuts	How 4: Material / Alloy	How 5: Power Drive (Belt & Pulley)	How 6	Requirements Planning: Completeness	Current Internal Importance	Planned Internal Importance	Improvement Ratio	Absolute Adjusted Completeness	Relative Adjusted Completeness	Competitive Benchmark: ○ Bag Palletizer Gantry	△ Gudel Available Product	◇ Importance to Competitor 3
1	Material efficient (lt-wt)	5	H	L	L	H	-		100	1	1	1.00	100	15%	5	3	2
2	Working speed(4m/sec)	5	L	L	L	-	H		60	1	1	1.00	60	9%	5	3	3
3	Girder	5	H	M	L	H	-		110	1	1	1.00	110	16%	1	1	1
4	Runway beam	5	H	M	L	H	-		110	1	1	1.00	110	16%	1	1	1
5	Legs	5	H	M	L	H	-		110	1	1	1.00	110	16%	1	1	1
6	Load	3	M	L	M	M	-		30	1	1	1.00	30	4%	4	3	3
7	Max.Displacement	2	M	-	L	M	-		14	1	1	1.00	14	2%	1	1	1
8	Working Environment	1	-	-	-	-	M		3	1	1	1.00	3	0%	5	3	3
9	Cost	5	H	L	L	H	H		145	1	1	1.00	145	21%	4	5	5
10		1							0	1	1	1.00					

		1	2	3	4	5	6
Technical Weights	Absolute	245	63	41	240	93	0
	Relative	36%	9%	6%	35%	14%	0%
	Importance	5	2	2	5	3	0
Technical Targets	USL						
	Target						
	LSL						
	Units						

Fig.5.3.1 House of Quality for Bag Palletizer Gantry.

The above house of quality shows that relatively Beam selection is most important area followed by material property of that selected section. After completing beam and material areas the power drive effect on parameters of beam section.

6. CONCEPT GENERATION

This chapter will deal with concept for a new light-weight bag palletizer gantry system with purpose of having significant cost reduction comparing to current model.

6.1 GUDEL Current Bag Pattelizer Gantry Model:[6]

Before start to deal with new concept for future gantry, it is necessary to discuss the current model. The type of section and technical specification will be introduced briefly.

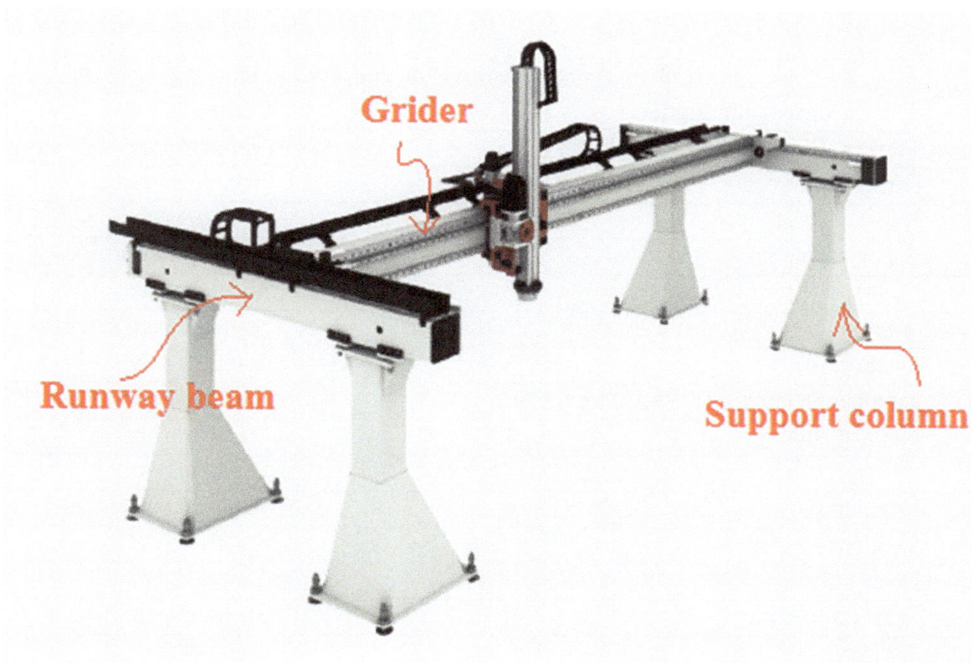

Fig6.1.1 GUDEL standard FP Gantry model.

The current solution uses the GUDEL standard FP model. Each part of this model is made of aluminum extrusion process. Drive system use in this gantry is Rack and pinion drive with GUDEL standard motor and all other drive assembly. Each of gantry parts are connected with nut and bolt joint.

The FP model is available in seven different configuration varies with payload and its size. The FP model is basically use for linear motion in all three direction.

Despite the gantry light-weight, provided by relative small section proportionally affected on the gantry cost. FP model having high weight due to extruded aluminum part and power system. The total cost of the model is also too high, and Indian customer not affords such expensive machine for bag palletization.

In order to improve the current model, the focus will be to create a model that can reduce weight but at the same time use alternative power drive system for high speed and light weight assembly.

6.2 The Concept Model

This section has the goal of study new configuration and evaluates their potential as new Bag palletizer gantry model .Therefore, the focus is the use of determinate section and study all load cases one the bases of requirement list. The power drive study is an important consideration before beginning of design study.

The design philosophy behind the proposal was the focus on displacement and stress value. In order to prevent deflection close to allowable limit, but also power drive also has better effect on light-weight aiming.

Table .6.2.1 List of component used in concept model.

List of component required in concept model	
Component Name	**Status**
Carriage Plate 595×425mm	Standard part
Girder Beam	Need to design
Runway Beam	Need to design
Support column	Need to design
	Standard part
	Standard part
Guide way	Standard part
Lubrication unit (4 for each carriage plate)	Standard part
Nut and Bolt	Standard part
	Standard part
Mounting	Standard part
Lubrication unit	Standard part
Drive pin	Standard part
Worm gear unit	Standard part
Flange	Standard part
Spacer	Standard part
Coupling	Standard part
	Standard part

This above component table is referred from GUDEL standard FP gantry model. With the help of this study design of parts like Girder, Runway Beam and Support column is carried out .While design this parts various literature survey help to get proper design .

6.3 Design Phase 1

With help of quality index study and literature survey few assumptions are referred. The proposed model is designed having cross section study as follows. First of all design start with Girder section.

Girder is center part of Bag palletizer gantry, which is fix at both ends with carriage plate(fig FP model) The manipulator and gripper is attach to this girder part with carriage plate connections. The carriage plate is slide on Girder with help of Guide ways and roller arrangement, which moves carriage plate and hoist assembly in linear manner. The power drive use to move Carriage plate is of belt and pulley drive [Appendix 1] . For given manipulator load and Bag load with speed requirement is use in Belt and pulley design. With all this consideration put in to mind while design of Mild steed Girder.

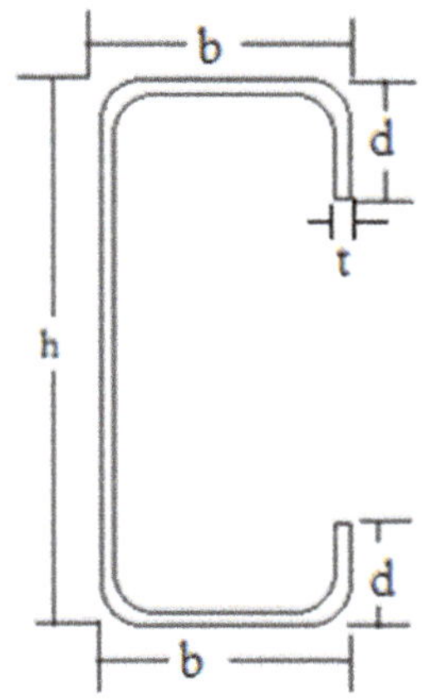

Fig.6.3.1 C-Section parameter

Table6.3.1 Paremeters of C-Secrion

Parameter	h	B	d	T
Value(mm)	210	110	33	3

With the help of Quality index study and literature survey few assumptions are referred .The proposed model is designed having cross section study as follows. First of all design start with Girder section.

Girder is Design with Basic engineering theorem of beam design. In this condition beam is fix at both ends, and concentrated load of manipulator is applied at center .This study gives us an few required values for further design. Now, the total load applied on girder is weight of hoist assembly which is dead load and weight of Bag which is live load .

Given Data :

Weight of bag : 981N

Weighty of Hoist assembly:1947.79 [6]

Length :2.4M

Total applied load (P)=2928.79N (wt.of bag + wt of Hoist assembly)

This project explains the theory of ultimate load applied to a Beam fixed at both end and simply supported beam based on the concept of ultimate load for beam of C-Section cross-section under pure bending. Pure bending refers to flexure of a beam under a constant bending moment. Therefore, pure bending occurs only in regions of a beam where the shear force is zero. In the simplified engineering theory of bending, we make the following assumptions:

1. The beams are assumed to internally statically indeterminate.
2. The strains caused by the deformations have a relationship with stresses.

3. When Mp is reached, a plastic hinge is formed.

The shape factor gives a very good estimate as to how much the yield moment M_y, could be exceeded before the ultimate plastic capacity is reached. Basically, shape factor can be calculated by this simple equation:

Shape factor $= \frac{M_{Ult}}{M_{Yeild}}$

$M_{Yeild} = \frac{\sigma_y I}{C}$

For the beam experience plastic yielding, the above equation will be altered to :

$M_{Ult} = \sigma_y Z$

$Z = \frac{A(x+y)}{2}$

A= area of cross section, x&y = distance of load to natural axis.

Case 1: C-Section parameter,

Moment of inertia :I = 10.235 x E^6 mm^4

Given that the yield strength of Cold rolled mild steel $\sigma_y = 370MPa$.

M_{Yeild} = 36.1KN-m

M_{ult} = 115.6 KN-m

Shape factor =3.2

For Girder Beam maximum moment

$M_{ult} = \frac{PL}{8}$,[ ]

P_{ult}=385KN

P_{yeild}=120KN

For Runway Beam maximum moment

$M_{ult} = \frac{PL}{4}$,[ ]

P_{ult}=191KN

P_{yeild}=59KN

The above calculation shows that, for girder beam the yield load is 120KN and yield occurs in runway beam at 59N.

1) Check if given beam section satisfies yield load value
2) Power drive assembly is also main consideration while section selection.[6,8,12]
3) Few literature will help to get proper section on particular load condition[2]

Case 2: Design for support column

In this case of column with fixed at one end and free at other end comfiguration subjected to compressive load as shown in fig., the deflection pattern is something like that shown in fig..

Let us take an arbitrary section a-a at distance x from fixed end A of the column.If we take moment about C, we can write

$$M_x = P_y - P\delta$$

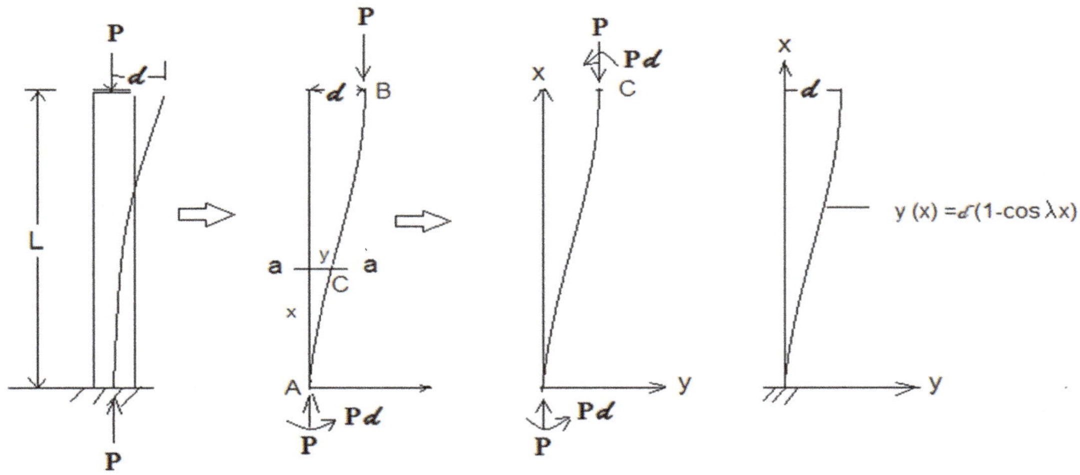

Fig 6.3.2 Support Column with Fixed-Free condition

Combining it with,

$EI\frac{d^2y}{dx^2} = - M_x$

Simplify above equation ,substitute λ^2=P/EI

The solution of the above equation is,

$y(x)=A_1 \cos(\lambda x) + A_2 \sin(\lambda x)+\delta$..(eq.1)

Applying the boundary condition, at fixed end x=0

Hence, the equation becomes

$y(x)= \delta(1-\cos \lambda x)$..(eq.2)

This is the equation of the deflection curve .it may be noted that at the free end,lateral deflection is δ. That is y(L)=δ

So, $\cos(\lambda L) = 0$

Minimal value of λL satisfying this condition is

$\lambda L = \frac{\pi}{2}$

$\lambda^2 = \frac{\pi^2}{4L^2}$

So, critical load for column is given by expression(by putting λ^2=P/EI)

$P_{cr} = \frac{\pi^2 EI}{4L^2}$

Calculation:

Load on each column =1468N (wt. of girder and wt. of runway beam)

E(M.S)=205GPa

Length of column=1800mm

$I = 9407mm^4$

So, to avoid buckling for applied load of 1468N the moment of inertia of section must be greater than $9407mm^4$.

7. GEOMETRY OF BAG PALLETIZER GANTRY MACHINE.

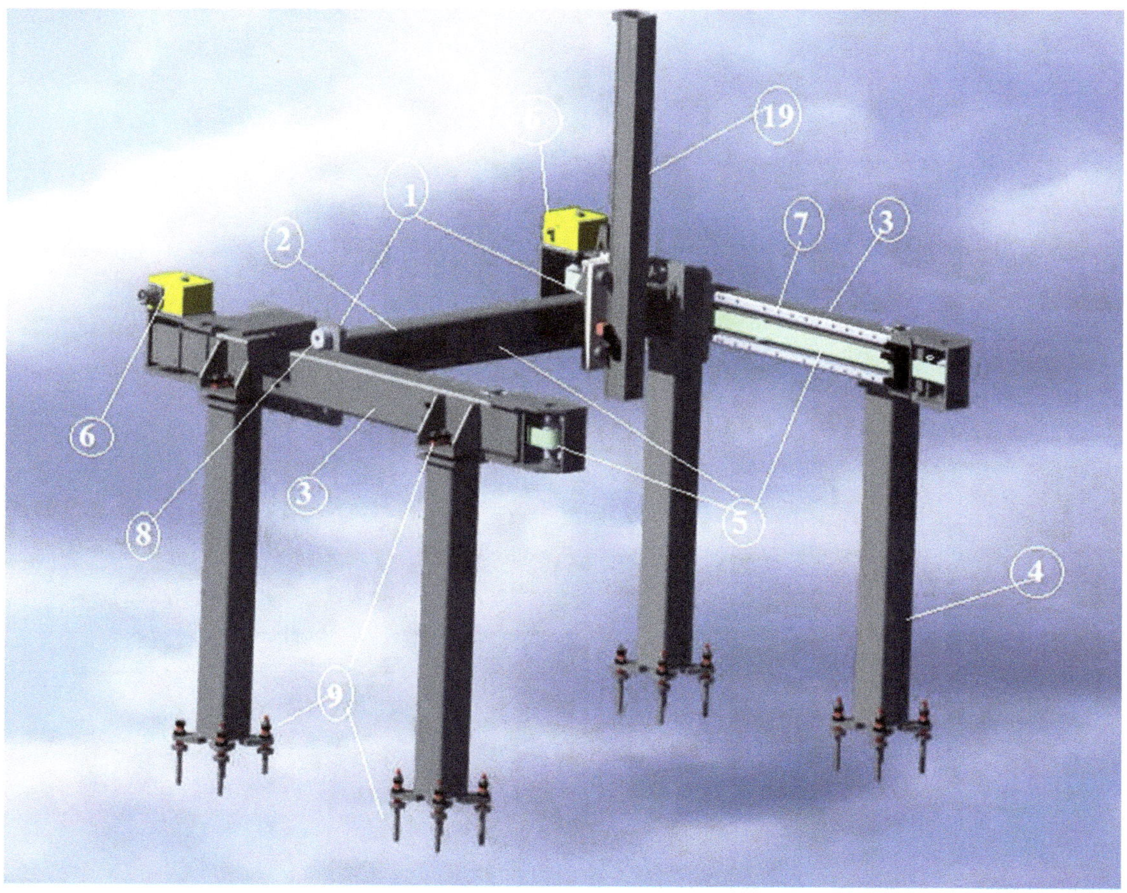

Fig.7.1.1 Bag Palletizer Gantry Model.

The bag palletizer gantry can be divided into three main parts .They are named as Hoisting system, Girder system, Runway system. Each system has particularly performed same task at its position to move linear along the line. On the other hand, Bag palletization system provide picking and placing action from conveyer to palletizer .In this Bag Palletizer Gantry model the main design criteria is to consider the drive system and its assembly. Because, that drive system must be fitted in the selected section. The standard Gantry model (FP) is driven by rack and pinion drive system. In that model design criteria is based on rack and pinion drive assembly and its size. Considering this the new Bag Palletizer Gantry model is to be designed. The BPG is designed for customer requirement, and every modification has based on standard model of gantry system.

The numbers of component used in BPG are based on standard model parts. Following table considering total number of component used in BPG model.

Table 7.1.1 List of component of BPG

em Numbe	Part Name	Quantity	Status
1	Carriage Plate	3	Standard part
2	Girder Beam	1	Need to design
3	Runway Beam	2	Need to design
4	Support column	4	Need to design
5	Belt	3	Standard part
6	Motor	4	Standard part
7	Guide way	12	Standard part
8	Lubrication unit (4 for each carriage plate)	16	Standard part
9	Nut and Bolt	246	Standard part
10	Roller	10	Standard part
11	Mounting	10	Standard part
12	Drive pin	3	Standard part
13	Worm gear unit	3	Standard part
14	Flange	3	Standard part
15	Spacer	3	Standard part
16	Coupling	3	Standard part
17	Pulley	6	Standard part
18	Bearing for pulley	12	Standard part
19	Manipulator	1	Standard part

This section summaries all the different parts of Bag palletizer gantry as well as the different stage of production that are required. But firstly give a better understanding of the parts position in global structure. It is going to present a figure containing a frontal, lateral and perspective view of the Bag palletizer gantry.

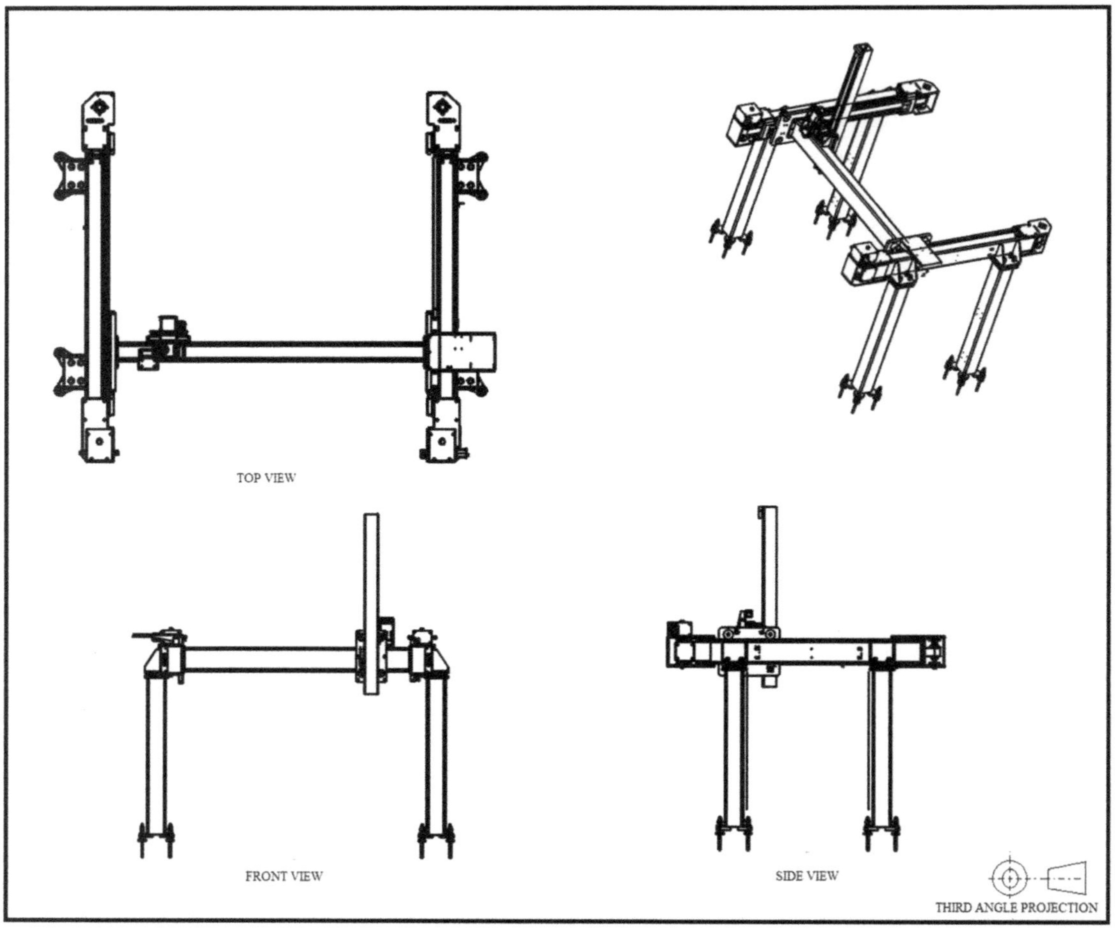

Fig 7.1.2 2D model of Front, Top, Side and Perspective view.

7.1 Design strategy

This section addresses the design strategy for the proposed Bag palletizer gantry system. The design strategy divides the tal amount of the Bag palletizer gantry assembly's parts into four categories, where each contemplates a major type of part. The referred categories are: strategic parts, core parts, purchased parts and finally consumables.

Strategic parts are related to variations of the main product, in this case a Bag palletizer gantry. It is consider that this field related to special orders .This part consider modification toward customers requirement and at that time customer is happy to pay extra cost for extra variation in Given product. The core part related to cold rolled beam element that consists of the structural parts.

In fig7.1.1 Parts of Bag palletizer gantry system are shown with their parameters. In this parts manipulator are attached to girder beam with attachment of guide ways and roller with carriage plate. The Girder beam is attach to runway beam with carriage plate and roller. but one side with two plane roller and other end of girder is attach with four grooved roller. This is because free motion of girder without perfect restriction in motion. The runway beam are two side of girder end with identical size and same power mechanism are attached. The end locations of runway beam are attached with support plates to guide the perfect assembly of motor and pulley system.

Table 7.1.2 Dimension details of Hoist assembly parts.

Element	Parameter	Value
Hoist beam	Length	125
	Width	125
	Height	1200
	Thickness	10
Carriage plate	Length	595
	Width	425
	Thickness	30
Roller	Diameter	10
Guide ways	Length	1200
	Width	24.7
	Height	33
Motor	Length	200
	Width	100
	Height	100

Table 7.1.3 Dimension details of Girder mechanism.

Element	Parameter	Value
Girder beam	Length	2400
	Width	110
	Height	210
	Thickness	6
End plate	Width	250
	Height	250
	Thickness	12
Belt	Length	3400
	Width	85
	Thickness	5
Pulley	Outer diameter	100
	Inner diameter	40
	Length	100
Pulley shaft	Diameter	40
	Length	110

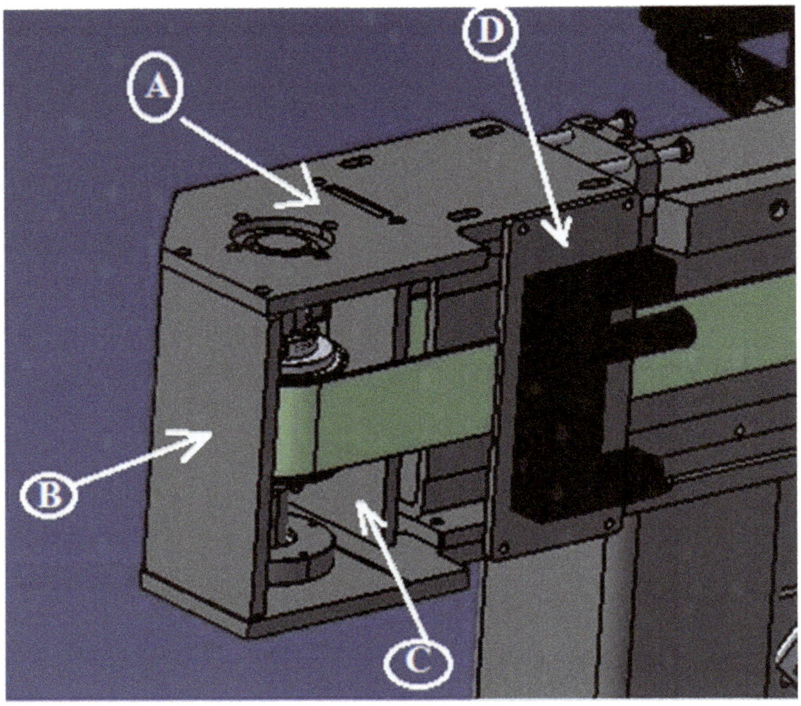

Fig 7.1.2 Runway Beam Assembly Parts

Table 7.1.4 Dimension details of Runway beam assembly

Element	Parameter	Value
Runway beam	Height	210
	Width	110
	Thickness	6
	Length	2400
Plate A	Length	400
	Width	150
	Thickness	4
Plate B	Length	210
	Width	110
	Thickness	4
Plate C	Length	210
	Width	95
	Thickness	4
	Length	210

Plate D	Width	75
	Thickness	4
Pulley adjustment	Length	65
	Width	15
Pulley adjustment nuts	Size	
Guide ways	Length	1200
	Thickness	24.7
	Width	33

Fig 7.1.3 Support Column Assembly Parts

Table 7.1.5 Dimension details of Support column parts.

Element	Parameter	Value
Support column	Length	1800
	Width	200
	Height	200
	Thickness	3,4,6
Top end plate	Length	350
	Width	350
	Thickness	8
Plate E	Length	200
	Width	210
	Thickness	4
Plate F	Length	400
	Width	200
	Thickness	4

8. BAG PALLETIZER SIMULATION

8.1 Formation Specification

In order to effective estimate of bag palletizer gantry into the formation , it is essential to find the deflection and stress induced in BPG parts .Bag palletizer gantry is a frame structure and due to its working condition the analysis is an essential part of this thesis study. since, the bag palletizer gantry shows similar principle with Beam theory, and this theory was used to calculate the deflection and stress in the thesis study.

Deflection and stress calculation for BPG parts:

A) Main girder calculation are performed considering the influence of constant loading.

Constant loading are main girder weight as well as weight of all components connected to the girder such as guide ways, carriage plate, bearing, lubrication unit, motor, belt, pulley etc.

Loading evaluation

Main girder weight

$P_M = P_{manipulator} + P_{Carriage} + P_{DriveSys} + P_{assemble\ comp}$

$P_{manipulator} = 1640N$

$P_{Carriage} = 588N$

$P_{DriveSys} = 49N$

$P_{assemble\ comp} = 650N$

$P_M = 2927N$

Girder fix-end beam AB is loaded by a force P action at point D, Determine reaction at end point and deflection at D. Euler-Bernoullis beam theory of Beam fixed at both end has been used for analytical simulation of Girder beam.

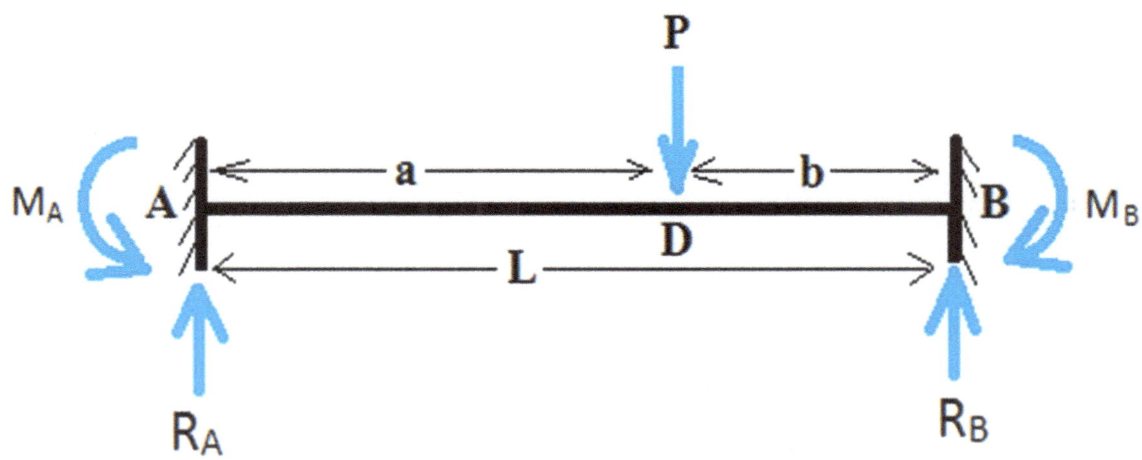

Fig.8.1.1 FBD of Girder

This is 2-degree of indeterminacy problem select M_A and M_B at the redundant

$R_A = \frac{Pb}{L} + \frac{MA}{L} - \frac{MB}{L}$

$R_B = \frac{Pa}{L} - \frac{MA}{L} + \frac{MB}{L}$

Thus , $\delta_D = \frac{Pa^3b^3}{3L^2EI}$

If $a = b = \frac{L}{2}$

Then bending moment , $M_A = M_B = \frac{PL}{8}$

Force reaction ,$R_A = R_B = \frac{P}{2}$

And deflection at center $\delta_c = \frac{PL^3}{192EI}$

Table 8.1.1 Parameter for Girder Calculation.

Sr.no	Parameter	
1	P_M (Main Girder Load)	
2	(Length of Span)	2 400mm
3	(moment of Inertia)	10.235×10^6 mm^4

Maximum deflection of Girder at center loading,

$$\delta_{max} = \frac{2928 \times 2400^3}{192 \times 205000 \times 10.235 \times 10^6}$$

δ_{max} =0.1004mm

B) Calculation for Runway beam:

Runway beam is supported by support column at two points so Euler-Bernoulli theory of simply supported beam with point load is used for analytical simulation of Runway beam. The Runway beam is supported at point A,B. and load P is applied at point D. The length of beam is Land load is applied at point a from support A.

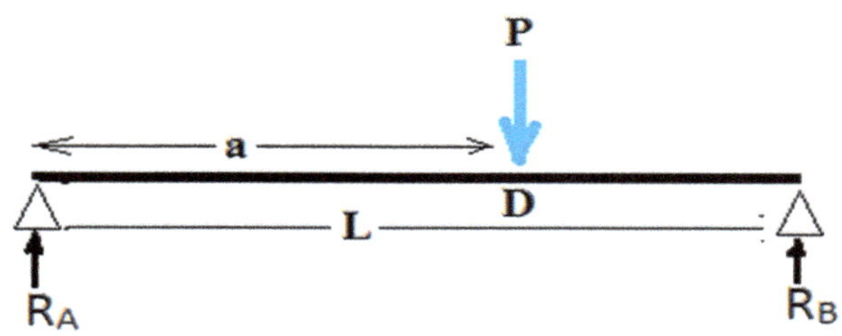

Fig.8.1.2 FBD of Runway Beam

Step 1: Ground reaction at support A and B

$R_A=(1-\frac{a}{L})P$, $R_B=\frac{Pa}{L}$

At x=L/2 , the deflection reaction the maximum as $\delta_{max} = -\frac{PL^3}{48EI}$

Table 8.1.2 Parameter for Runway Beam Calculation.

Sr. no	Parameter	Value
1	R_M (Main Runway Beam Load)	2032.5N
2	L (Length of Span)	1600mm
3	I (Moment of Inertia)	10.235×10^6 mm

The maximum deflection of Runway Beam at center :

4

$$\delta_{max} = \frac{2032.5 \times 1600^3}{48 \times 205000 \times 10.235 \times 10^6}$$

δ_{max} = 0.0826mm

8.2 Transferring the CAD model into the simulation environment

In the modeling and simulation of the model, two software's were used in the thesis study, Catia V5 and Hypermesh-Radioss.Catia was prefer to create the solid model of BPG machine. Geometries of the solid model were formed with the help of requirement from customer and available standard data of existing BPG model. Hypermes-Radioss (version v11) was chosen as a multifunctional simulation tool , in which Hypermesh used for mashing tool and Radioss for Post processing of simulation. During the transfer, all database of the model which includes the geometry of the parts , relationship of parts in the assembly, appearance properties are conveyed to Hypermesh software,with the help of standard [.step] file format .

A Bag palletizer gantry mechanism combination includes lots of connection points in the assembly and curved surfaces on the parts. This situation causes complexities and inaccuracies in the simulation and takes long time for simulation. To eliminate the possible inaccuracies due to the irregular shape of the model ,to decrease the simulation time and focus on main interaction area, only beam parts and connection plates were simulated in Hypermesh –Radioss. The motar, pulley, roller, lubrication unit, bearing was not included in the simulation .

Moreover, before starting the simulation, some modifications were made on the main BPG model. All solid surfaces were converted in to its mid-surface. By this way, it was aimed to avoid possible bad mesh configuration on curved part. since, bad mesh configurations and the resultant errors on finite element can lead to undesirable simulation results.

8.3 Material property:

Material and Property module in hypermesh radios software is used to identify specification of different materials in the model. Materials can be characterized according to their behaviors such as mechanical thermal, acoustic etc. In this thesis study, materials were defined as elastic-perfectly plastic. In elastic perfectly plastic material, strain remain constant after yield stress (fig) To introduce this specification to the program, density as a general property, elasticity and poissons ratio were committed input data .

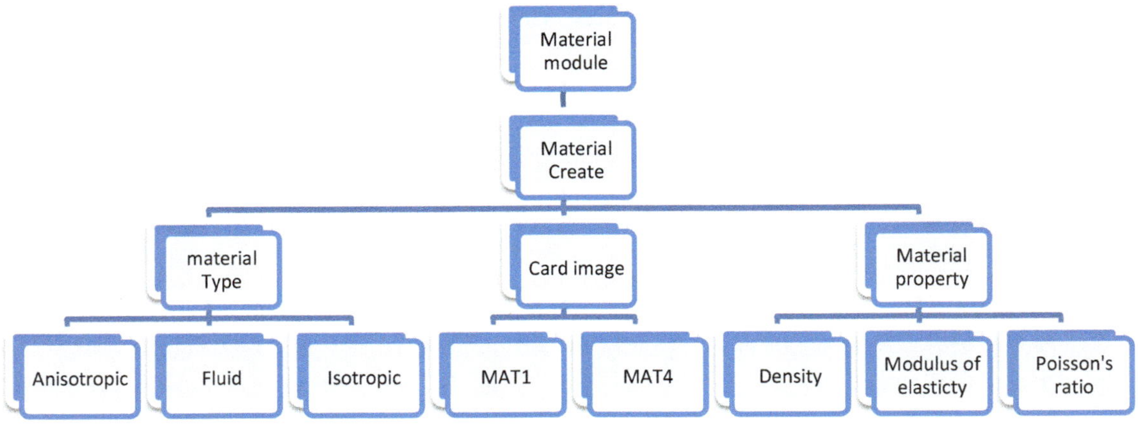

Fig 8.3.1 Hypermesh material module steps

this simulation, two different type of material were used to identify the material properties .mild steel alloy 1086 and result compare with aluminium alloy. After selection of material the actual distribution of material to particular part is done with material icon.

Table 8.3.1 Material Property

In

Material	Density	Young's Modulus	Poisson's Ratio	Yield stress
Mild Steel(1018)	7.87g/cm^3	205GPa	0.3	370MPa

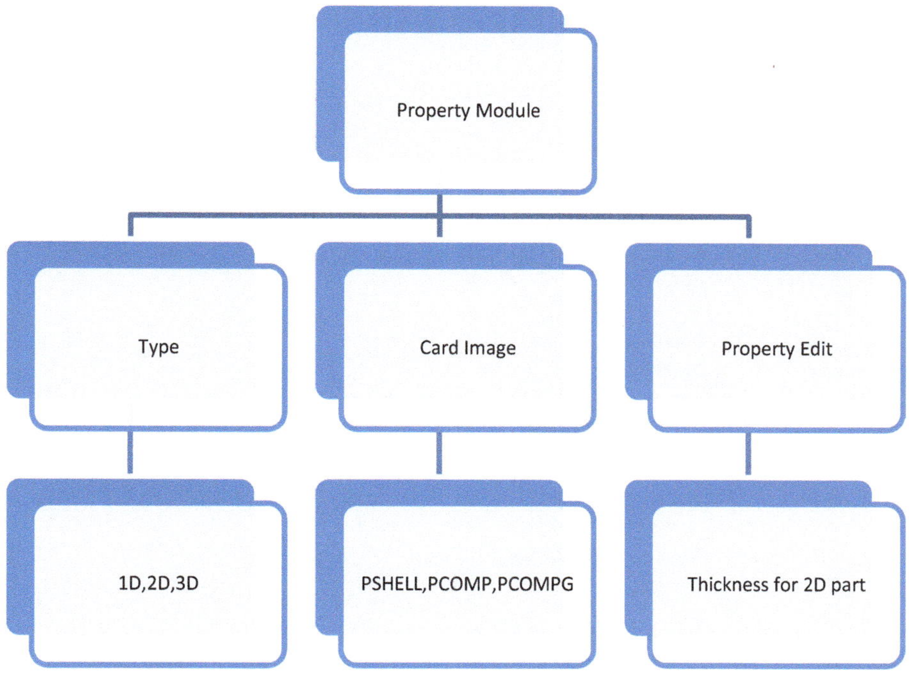

Fig.8.3.2 Hypermesh property module steps

8.4 Element type selection and creating mesh pattern:

Mesh is one of the most important factor of the simulation since accuracy and efficiency of the simulation depend on mesh quality of model parts .

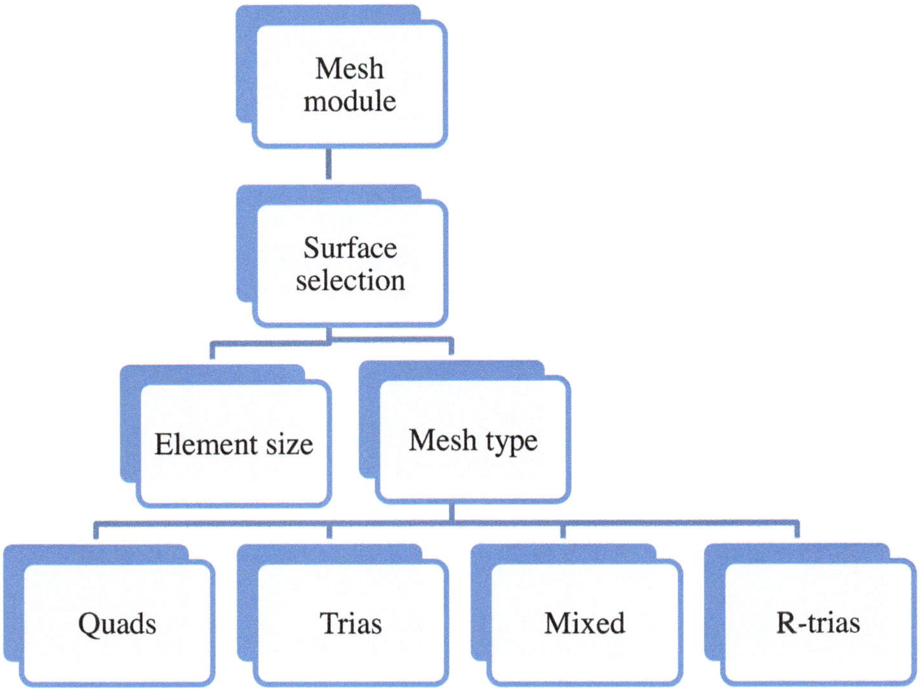

Fig 8.4.1.Hypermesh mesh module

It was avoided to create direct mesh on the model. Some part of assembly were partitioned to make more sensitive analysis on the particular place. The stress induced in particular part at certain load condition gives failure location on the model.

After defining nodes on all edges of the model, it is required to determine appropriate element type which connect the nodes and create two dimensional mesh networks. Through the element types, elements are the general two dimensional shell element of Hypermesh as shown in fig. The element can be used in a shape of quadrilateral and triangular. In the study, 4-node linear quad element was used due to its flexibility on 2D element with exact result.

Fig 8.4.2 Mesh model of Bag Palletizer Gantry

The quality index of hypermesh is an important part .Different quality parameters like skew, aspect ratio, jacobian are to measure of how far the given element deviates from ideal shape. Square means all angle 90 degree and equal side, while equilateral triangle is all angles 60degree and equal side. To reduce solution time element are mapped to local co-ordinate system instead of using a one co-ordinate system. Following table contain standard value for every quality cheack for quality result.

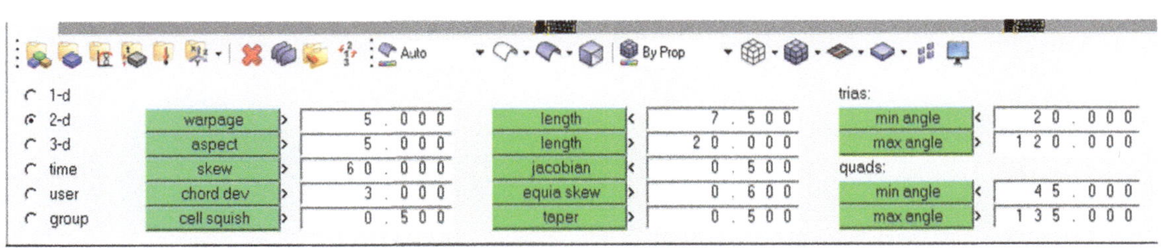

Fig 8.4.3 Quality check condition for Bag Palletizer Gantry mesh model

8.5 Loading and boundary condition:

In Hypermesh ,load and boundary condition can apply through tool window .Load can be created in the form of concentrated force. Besides, boundary conditions can be determined to assign the degree of freedom for each element in terms of symmetry and displacement.

In the thesis research, three case are studied for deflection and stress investigation of the Bag palletizer gantry were carried out. The study focus on compressive loading condition on beam section. Boundary condition and mesh properties remain same in all the simulation. Only section thickness property changed in simulation environment. Load and boundary condition are given below.

Fig.8.5.1 Model of Boundary condition.

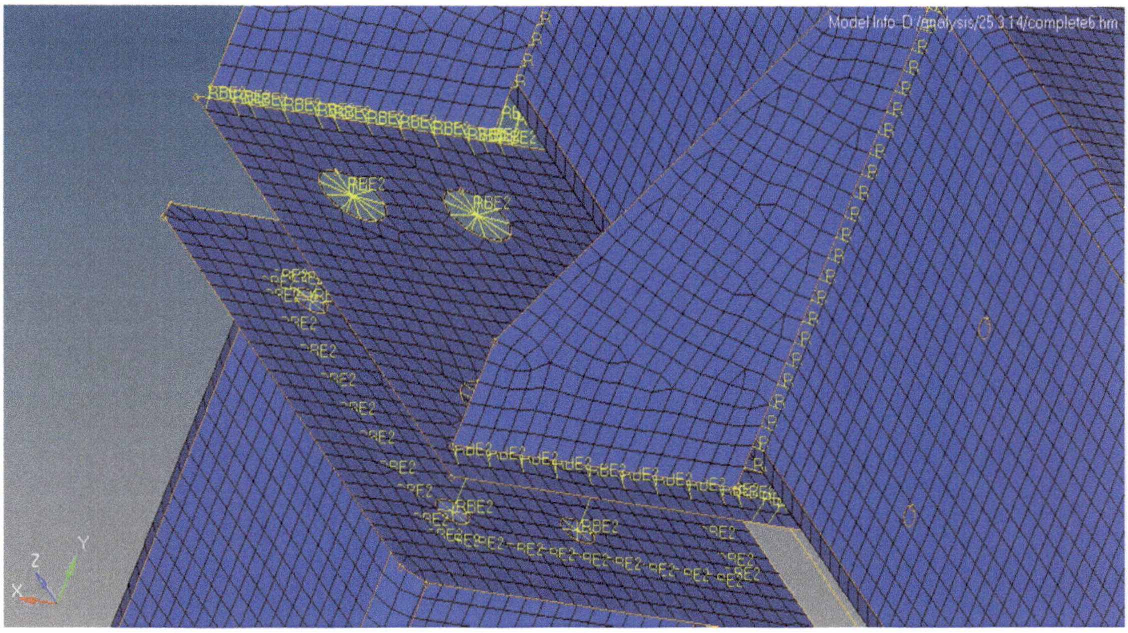

Fig.8.5.2 Rigid element connectivity.

In both the cases, the model legs are constrained and load 100kg is applied at manipulator end. The legs constrain condition all degree of freedom are locked. The connection of weld joint are attach with rigid element and nut and bolt are connected with spider arrangement of rigid element. But, the only change in section thickness is carried out in every case first case with 3mm thick mild steel and then with 4mm and 6mm thick section. The results for every cases are to be simulate in Radioss post processor.

8.6 Case study

This case study is about cold rolled mild steel C-member .This cold rolled mild steel beam generally failed due to buckling. Deflection of mild steel cold rolled beam in bending remains a unaddressed problem .Further, adequate experimental data in bending of beam is unavailable. Therefore, a series of test on C-member in bending has been completed at the structural testing lab of Johns Hopkins University. In this case beam tested four point bending test .The experimental data is being used help to develop and validate an efficient and reliable design procedure for C- member in bending.

A structural finite element model analyzed by Hypermesh Radioss software. The analysis result was setup and verified with experimental result. The finite element model consider the geometric in perfection and material nonlinearity .The analysis is extended to border range of section demand, material property and boundary condition.

8.6.1Testing Setup

The basic testing setup is shown in figure .The span length of test sample is 16ft (5486mm).The test based on four point bending condition .It consist of pair of 18ft log C-section open facing opposite to each other. The inner face is separated with hot rolled tube section, 10× 7.5×6×.025 inch.

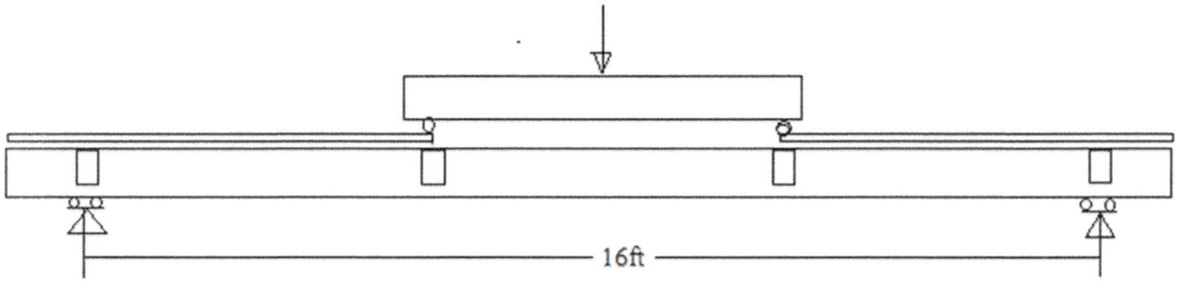

Fig.8.6.1 Elevation view of test setup

Fig.8.6.2 Panel setup for test[4]

Fig.8.6.3 Loading point configuration[4]

There is no panel placed in the constant moment region, remaining portion is attached to 0.019inch thick plate .This plate attached to the compression flanges in the shear spans only to restrict both the distortional and lateral buckling in this region. But, leaving only distortional buckling free region.

The loading system employed 20Kip (88964N) actuator. Which has a maximum 6 inch(152.4mm) stroke. The specimen deflection were measured at 1/3 point with position transducer.

8.6.2 Finite Element Modeling

The experimental results build a solid background for further finite element analysis that can help the buckling study to examine further cross section, different loading and support condition and other structural details used in today's industry. However, in order to simulate the real flexural bending test and to capture the buckling mechanism of cold rolled steel member finite element analysis is necessary.

8.6.3 Modeling Details And Boundary Conditions

The finite element modeling is shown in figure .The section selected for validation is D8C097-7E6W , Because, design specification of Bag Palletizer Gantry girder and runway beam is coincide with this sample .The selected cold rolled mild steel C – section having dimension which we simulate with software is given below.

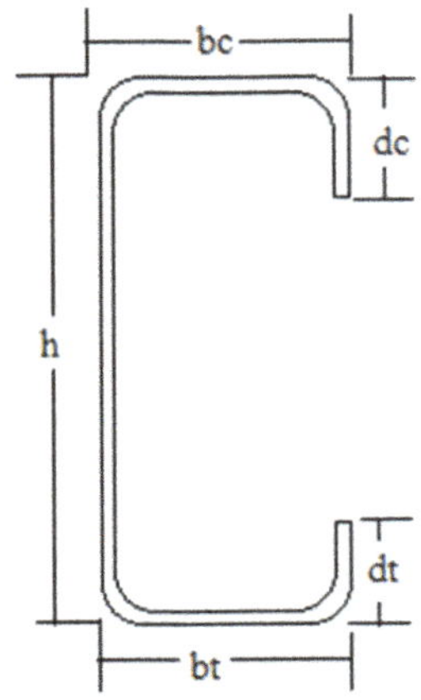

Fig.8.6.4 C-Section for test sample parameter.

Table 8.6.1 Design Parameter of test sample

Model dimension =D8C097-7E6W
h=206.5mm,bc=54.61mm,dc=16.51mm,bt=54.1mm,dt=15.74mm,t=2.54mm
E =203255N/mm^2 ,
Density =7.87 g/cc^2
Load = 2200N to 44400N
Yield strength=370Mpa
Poisson's ratio = 0.29

The beam arrangement is same as test setup. The beam is simply supported at the bottom flanges under two end tube. The tube is connected by bolt in real setup but that bolt are represented by 1D rigid element. These rigid elements are used to make rigid connection between tube and C- section attachment. Means to make displacement, rotation and all degree of freedom equal at two nodes. The same method is used for panel to upper section of C-member connection. The support point shows multi point constraints (MPC).

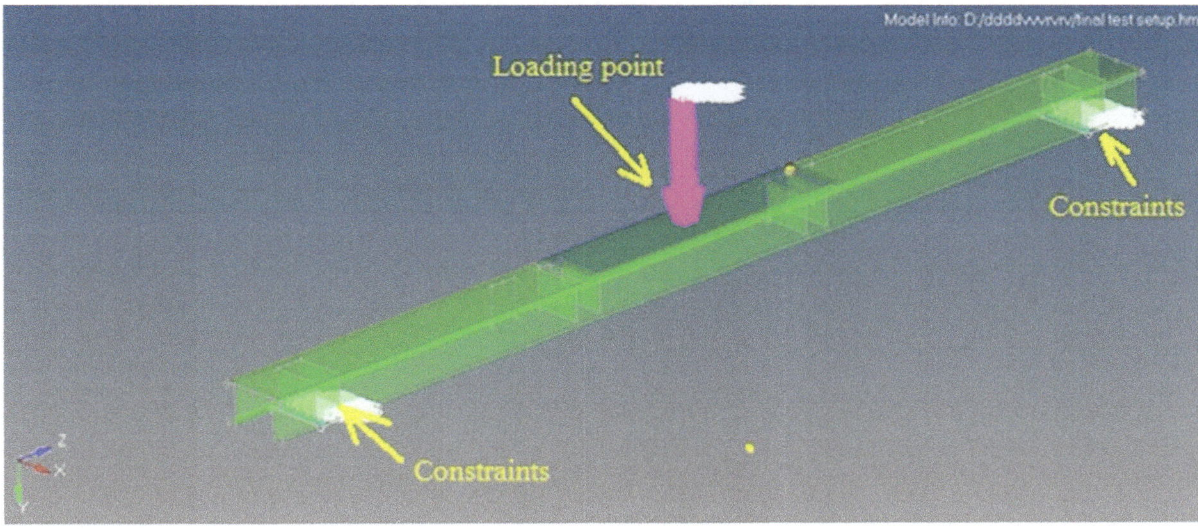

Fig.8.6.5 Finite element model.

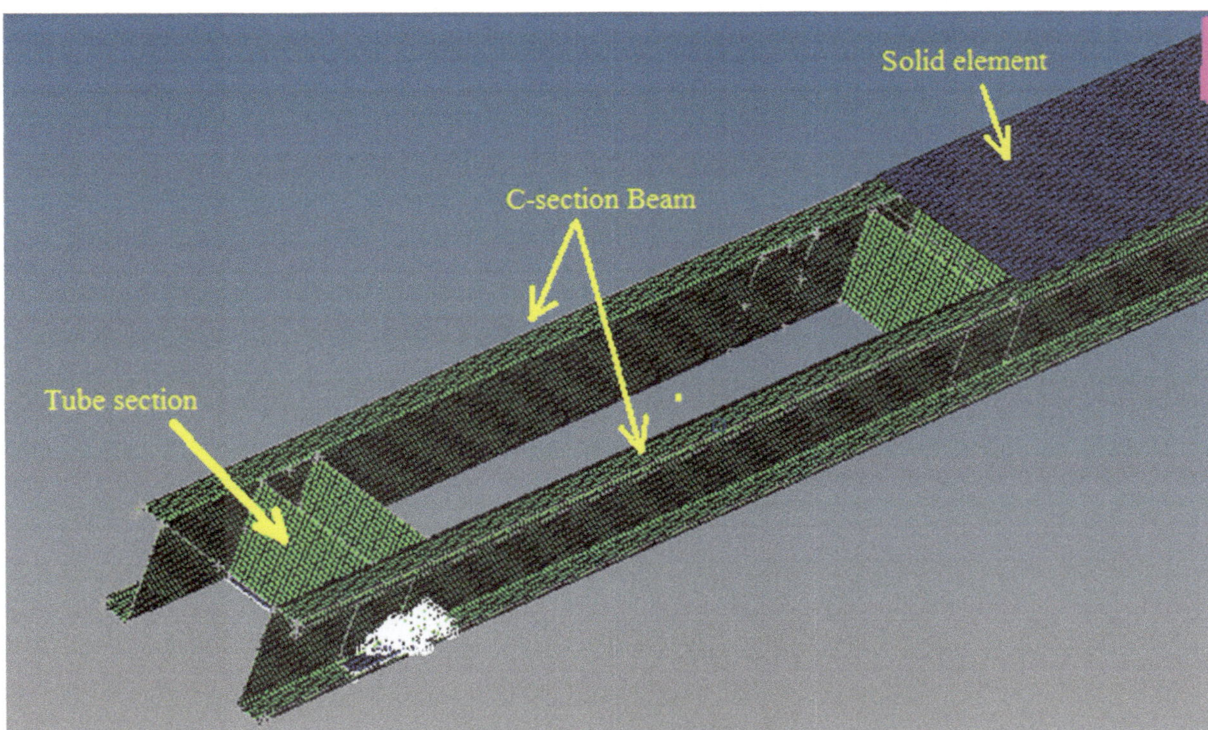

Fig.8.6.6 FEA model parts internal view.

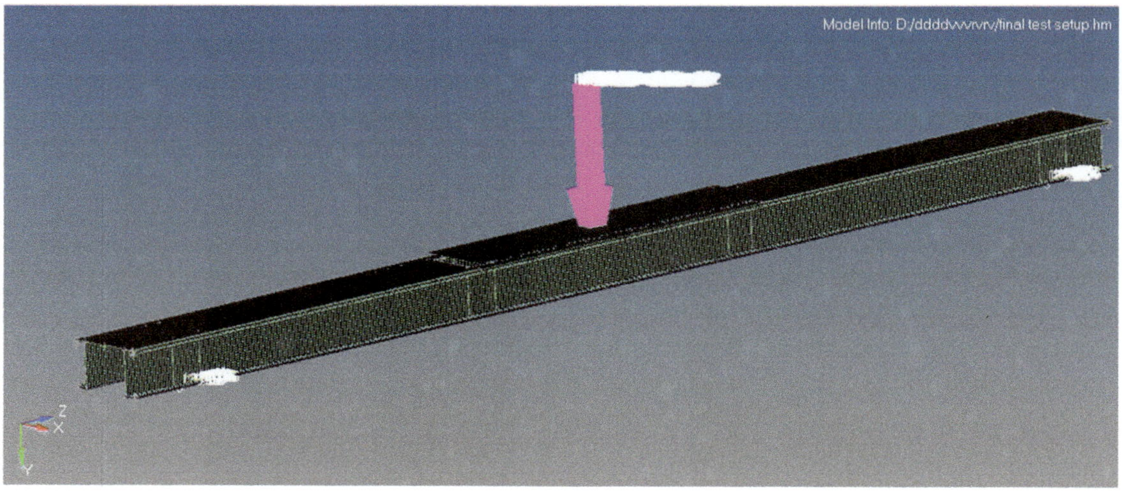

Fig 8.6.7 Mesh model of test sample

The analysis is carried out in Hypermesh Radioss with above material property and boundary condition. This analysis results are compare with test result to give validation of tested sample for given load condition. The test result and results obtain from Hypermesh Radioss analysis are discuss in next chapter.

9. RESULT AND DISCUSSION

Static analysis is the most common analysis method which is used in engineering.As the loads are assumed to be applied instantly,the effects due to the time variation are neglected.

9.1 Case 1 Stress and deflection investigation for 3mm thick C-Section:

This case is carried out in structural member of BPG.In this case load of 100kg is applied on manipulator end and result obtain from analytical calculation and numerical via software is comparatively studied. The variation on both the result is around 6% in girder deflection and 1.5% in girder stress value. This result contain stress and deflection value at various position on girder and runway beam.

Table 9.1.1. Problem size and job time summery

	Problem size	
Number of Element	Number of Node	Number of Component
146438	151382	120
Job Time Summery(Min)		
Load and Boundary condition	Material, Property and component	Stress and deflection analysis
13	16	26

9.1.1 Deflection on Girder :

Fig.9.1.1 HM Result of deflection BPG

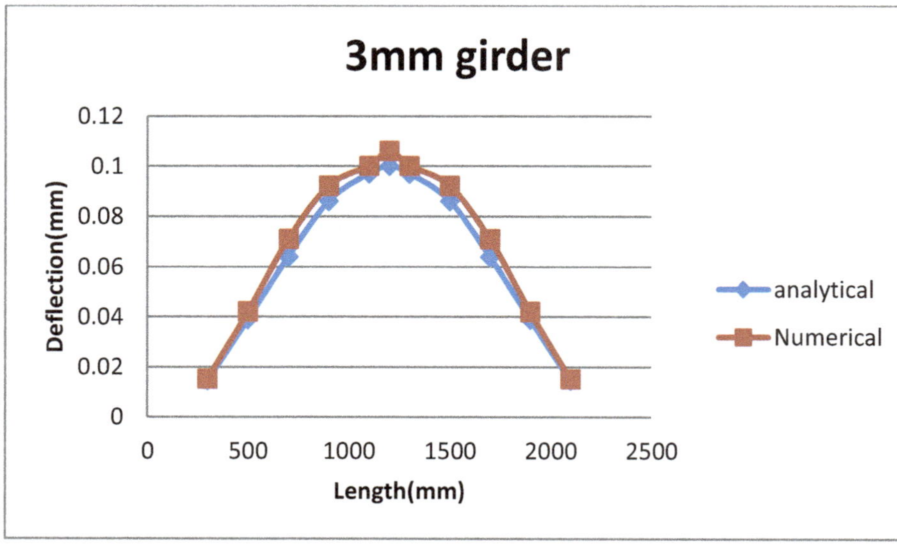

Fig.9.1.2 Graphical result of deflection of Girder beam

Table 9.1.2 Max. Deflection at center of girder

	Analytical	Numerical	% Deviation
Max. Deflection	0.1mm	0.106mm	6%

9.1.2 Stress Vs strain for Girder:

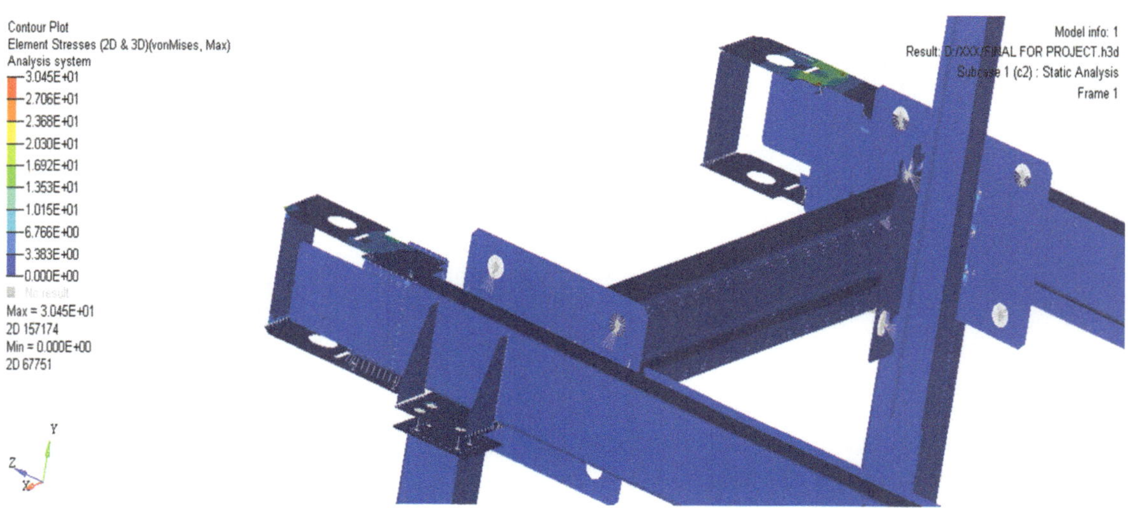

Fig.9.1.5 HM Result of stress in BPG.

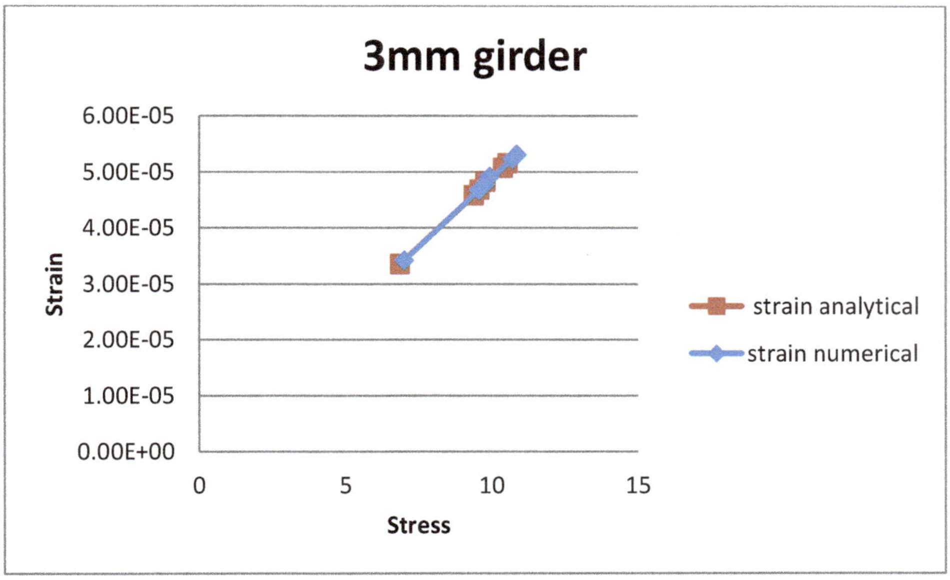

Fig.9.1.6 Graphical representation of stress vs strain of Girder Beam.

Table.9.1.4 Max. Stress at center of Girder Beam.

	Analytical	Numerical	% Deviation
Max.Stress	9.8mm	9.95mm	1.5%

9.1.3 Deflection of Runway Beam:

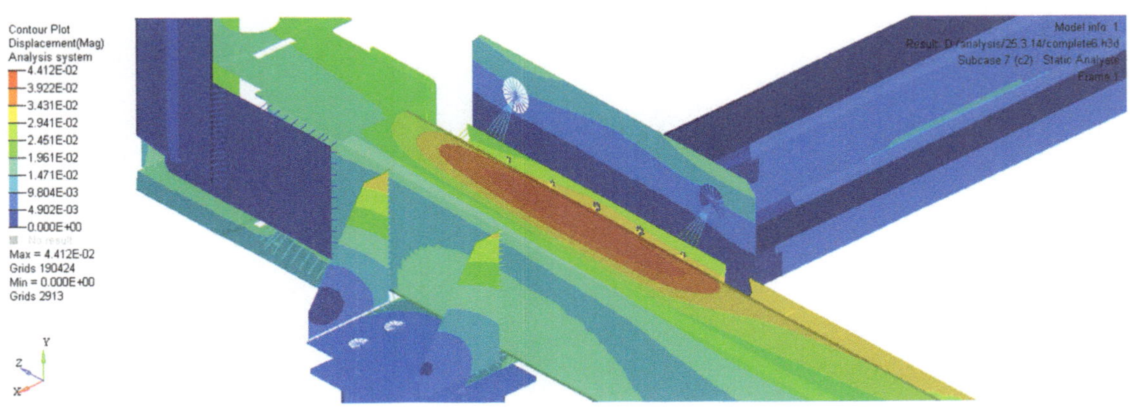

Fig.9.1.3 HM Result of deflection of Runway Beam

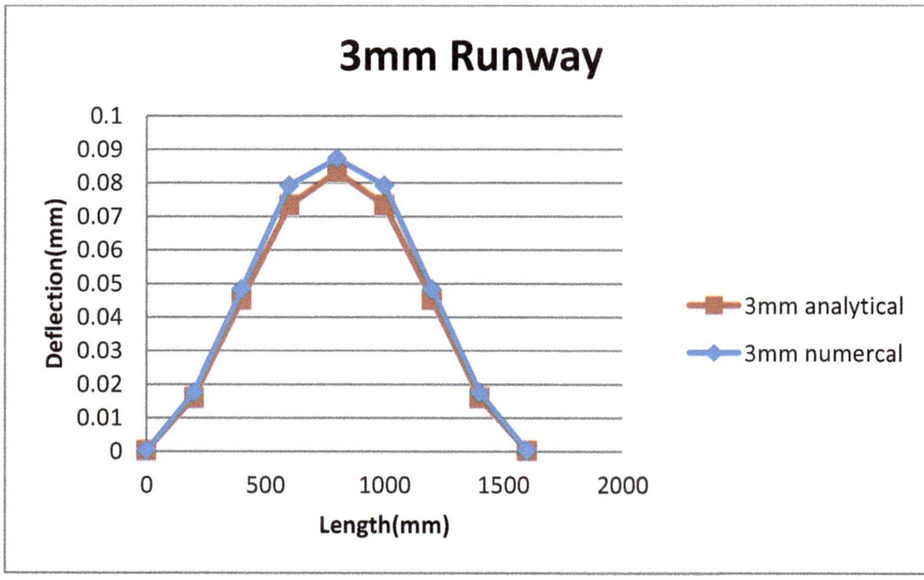

Fig.9.1.4 Graphical representation of deflection of Runway Beam.

Table .9.1.3 Max. Deflection of center of Runway Beam

	Analytical	Numerical	% Deviation
Max.Deflection	$0.083N/mm^2$	$0.087N/mm^2$	4.8%

9.1.4 Stress Vs Strain of Runway Beam:

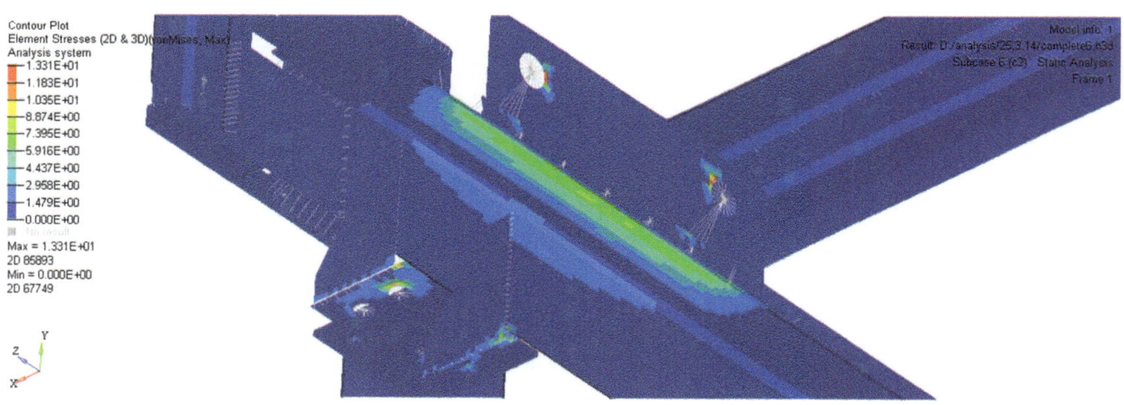

Fig.9.1.7 HM Result of Stress of Runway Beam.

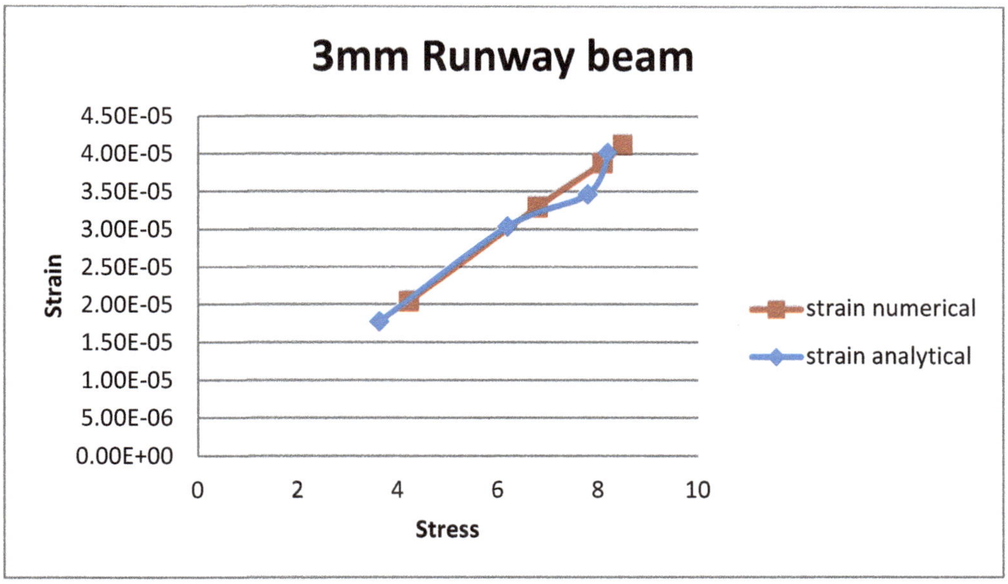

Fig.9.1.8 Graphical representation of stress Vs strain of Runway beam.

Table.9.1.5 Max. Stress at center of Runway beam.

	Analytical	Numerical	% Deviation
Max.Stress	8.2N/mm^2	8.5N/mm^2	3.6%

9.2 Case 2 : Stress and deflection investigation for 4mm thick C-Section:

The load and boundary condition are same as Case 1. But the only change is thickness of Girder and Runway member by 4mm. The difference on analytical and numerical calculation is discuss in following graph.

Table9.2.1. Problem size and job time summery

	Problem size	
Number of Element	Number of Node	Number of Component
146438	151382	120
Job Time Summery(Min)		
Load and Boundary condition	Material, Property and component	Stress and deflection analysis
13	5	29

9.2.1 Deflection of Girder Beam:

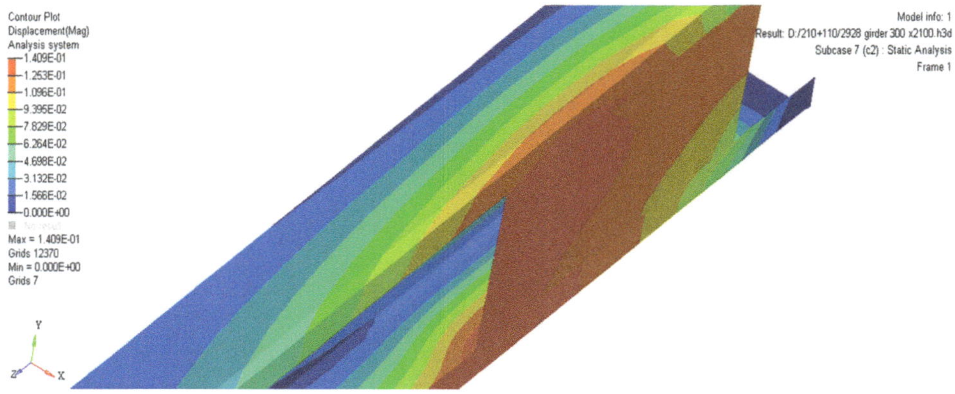

Fig.9.2.1 HM Result of Deflction

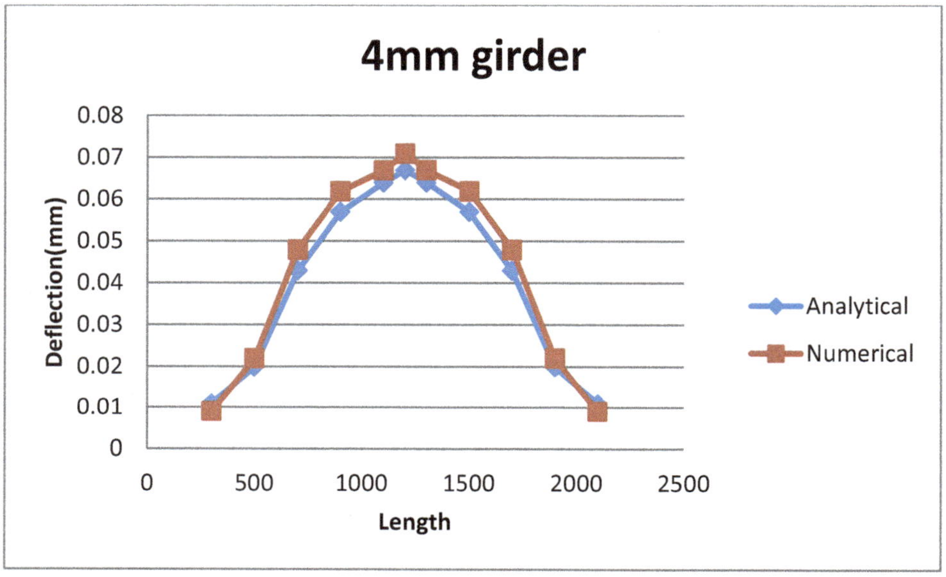

Fig.9.2.2 Graphical representation of Girder beam.

Table 9.2.2 Max. Deflection at center of Girder beam.

	Analytical	Numerical	% Deviation
Max.Deflection	0.067mm	0.071mm	6%

9.2.2 Stress vs strain of Girder Beam:

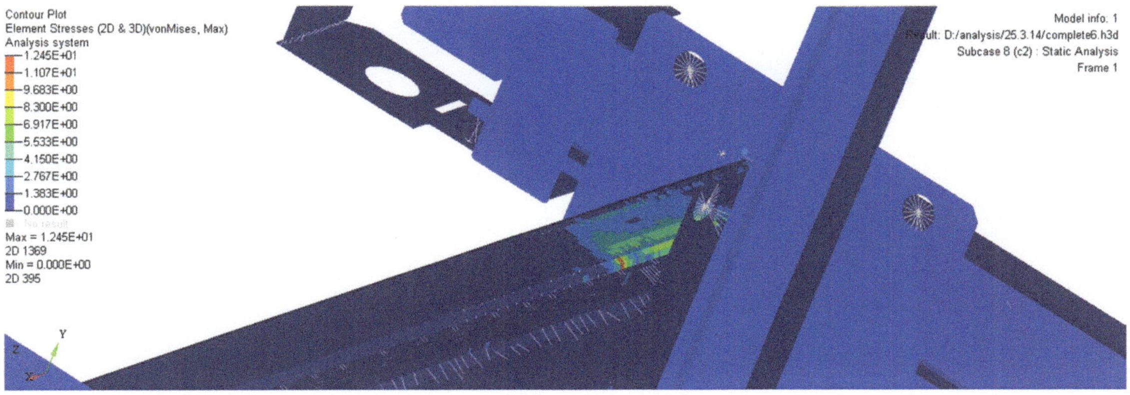

Fig.9.2.5 HM Result of stress

Fig.9.2.6 Graphical representation of stress vs strain of Girder.

Table 9.2.4 Max. Stress at center of Girder Beam.

	Analytical	Numerical	% Deviation
Max.Stress	6.52N/mm^2	6.74N/mm^2	3.3%

9.2.3 Deflection of Runway Beam:

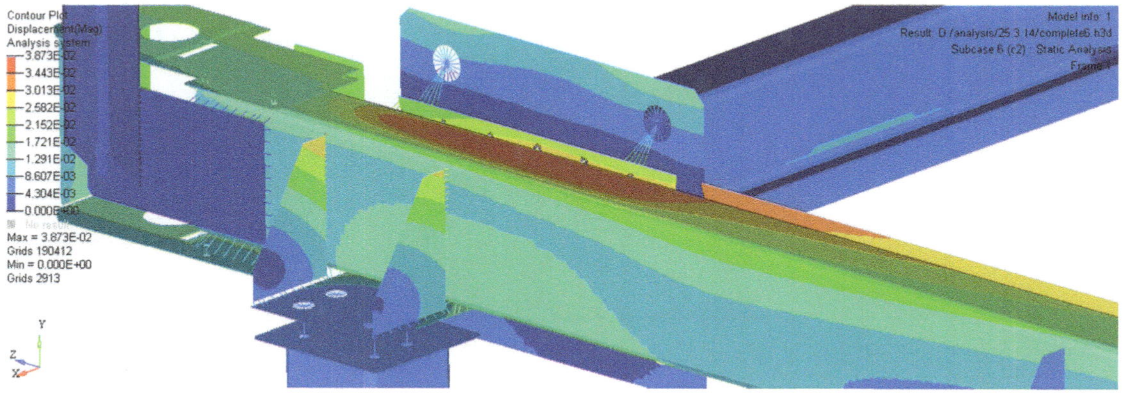

Fig 9.2.3 HM Result of Deflection of Runway beam.

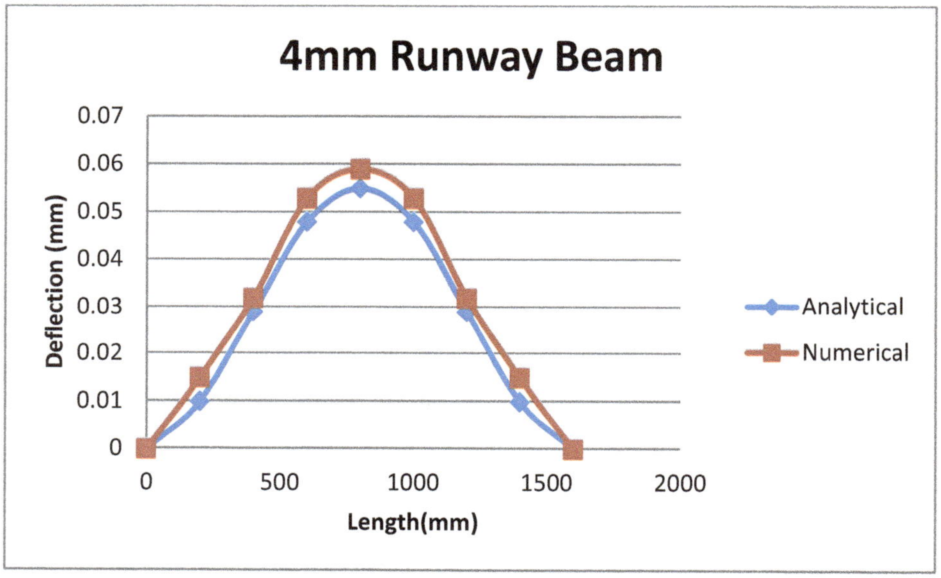

Fig.9.2.4 Graphical representation of deflection of Runway beam

Table 9.2.3 Max.Deflection at center of Runway beam.

	Analytical	Numerical	% Deviation
Max.Deflection	0.055mm	0.059mm	7.2%

9.2.4 Stress vs strain of Runway Beam:

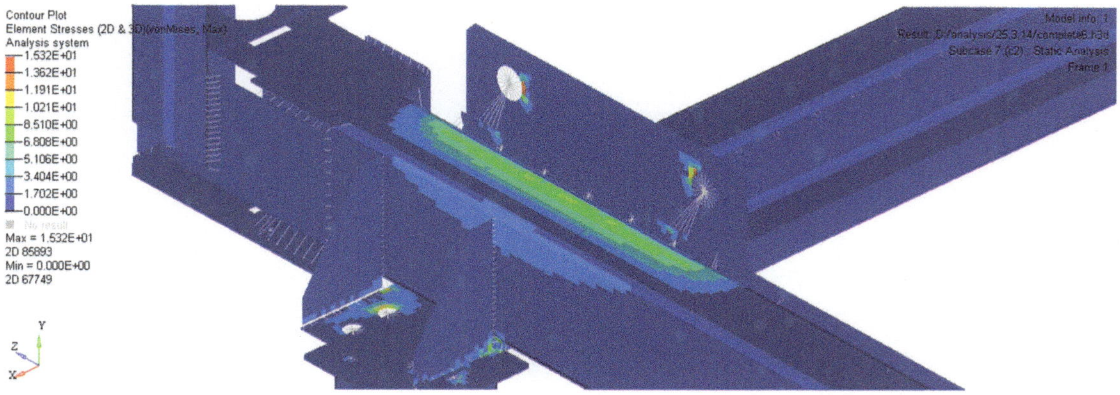

Fig.9.2.7 HM Result stress of Runway Beam.

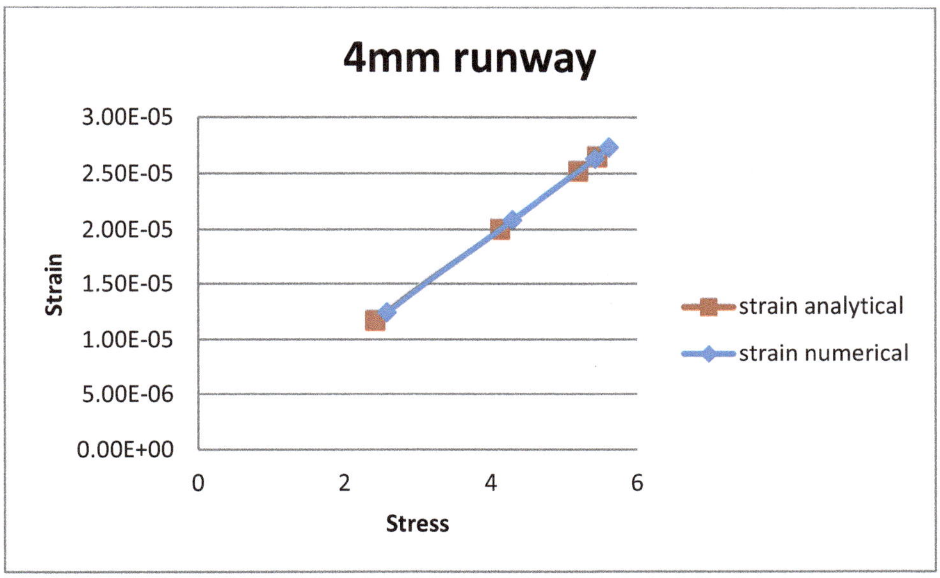

Fig 9.2.8 Graphical representation of stress Vs strain of Runway Beam.

Table.9.2.5 Max.Stress at center of Runway Beam.

	Analytical	Numerical	% Deviation
Max.Stress	5.46N/mm^2	5.62N/mm^2	2.9%

Case 3 Stress and Deflection study of 6mm thick C-section BPG:

Table.9.3.1 Problem size and job time summery.

	Problem size	
Number of Element	Number of Node	Number of Component
146438	151382	120
Job Time Summery(Min)		
Load and Boundary condition	Material, Property and component	Stress and deflection analysis
13	30	32

9.3.1 Deflection of Girder Beam

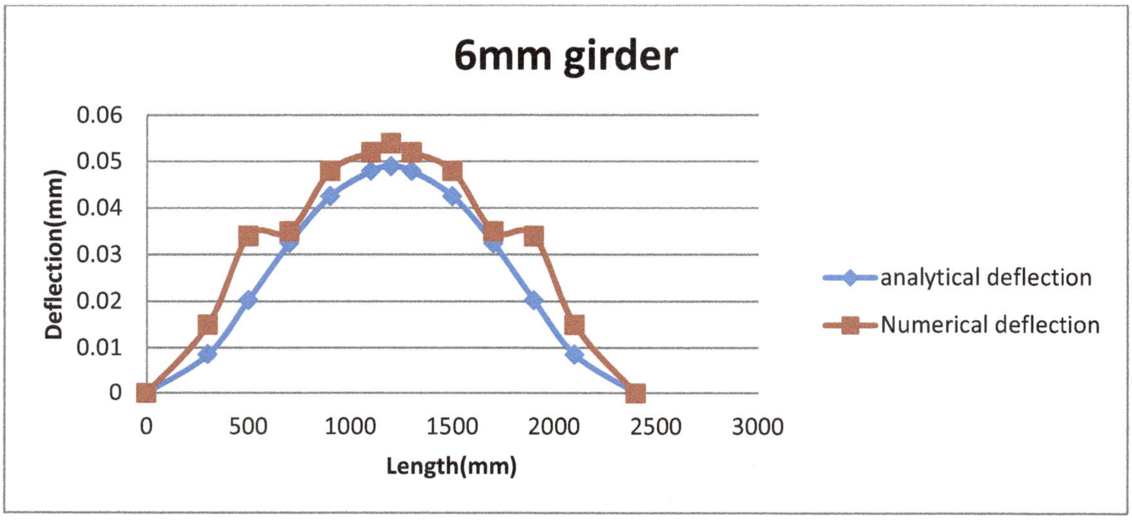

Fig.9.3.1 Graphical representation of Deflection of Girder beam.

Table 9.3.2 Max. Deflection of Girder Beam at center.

	Analytical	Numerical	% Deviation
Max.Deflection	0.049mm	0.054mm	10.2%

9.3.2 Stress vs Strain of Girder Beam

Fig.9.3.3 Graphical representation of stress vs strain of Girder

Table 9.3.4 Max. Stress at center of Girder Beam.

	Analytical	Numerical	% Deviation
Max.Stress	4.51N/mm^2	4.4N/mm^2	2.5%

9.3.3 Deflection of Runway Beam:

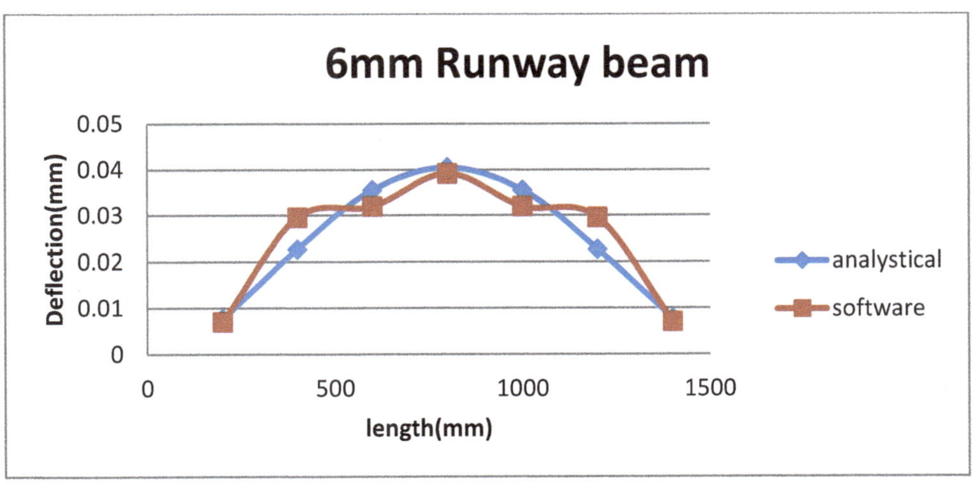

Fig.9.3.2 Graphical representation of Deflection of Runway Beam.

Table 9.3.3 Max.Deflection of Runway Beam.

	Analytical	Numerical	% Deviation
Max.Deflection	0.0404mm	0.0391mm	3.3%

9.3.4 Stress Vs Strain of Runway Beam

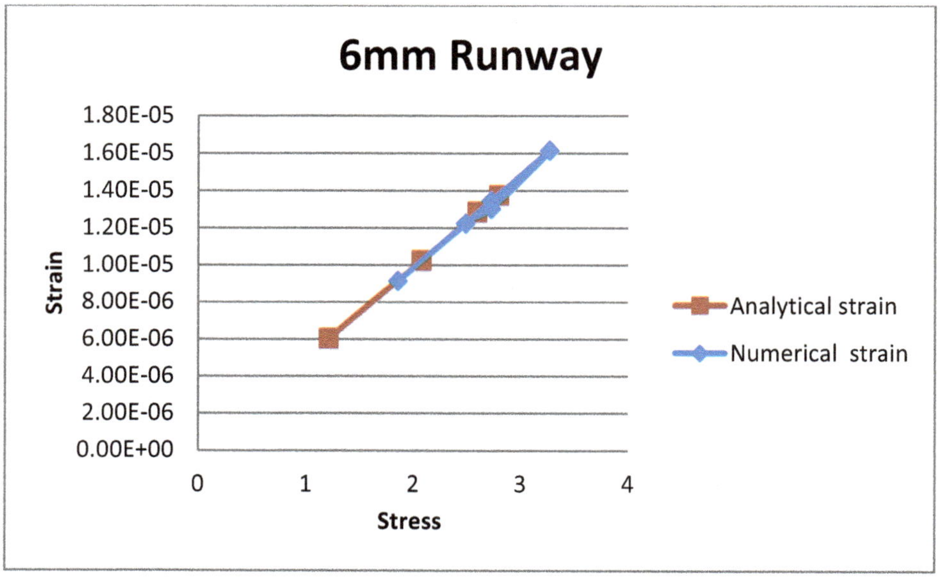

Fig 9.3.4 Graphical representation of stress vs strain of Runway

Table 9.3.5 Max. Stress at center of Runway Beam.

	Analytical	Numerical	% Deviation
Max.Stress	$2.8N/mm^2$	$3.27N/mm^2$	16%

9.4 Test Result

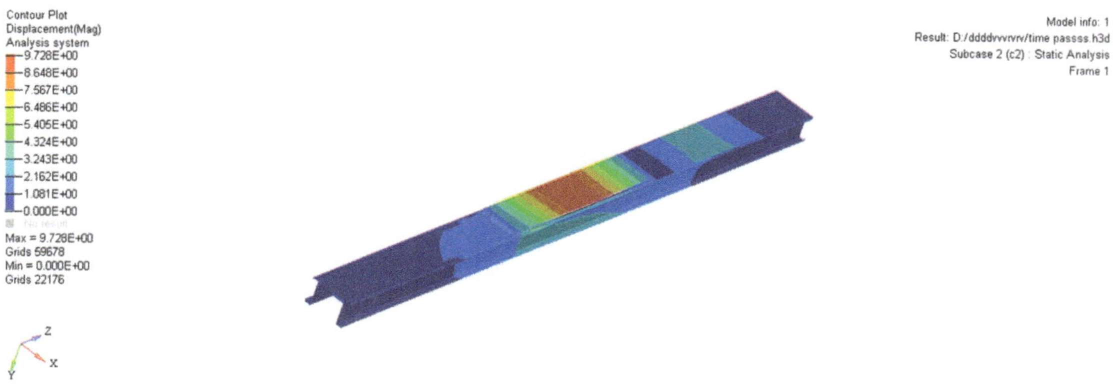

Fig.9.4.1 Test sample displacement at 8880N load

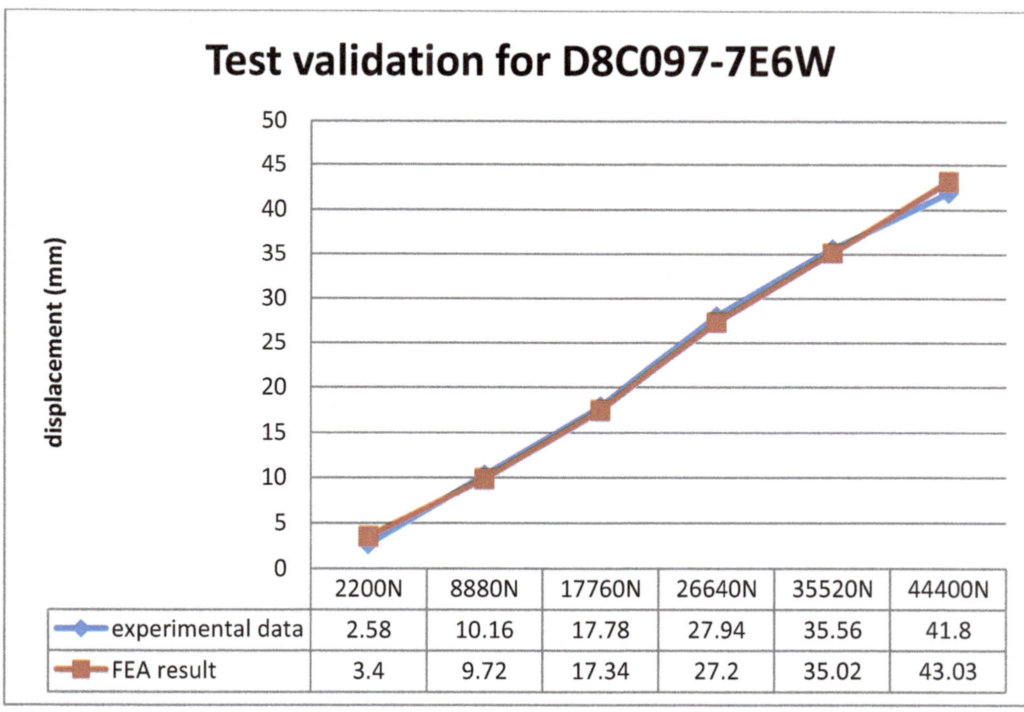

	2200N	8880N	17760N	26640N	35520N	44400N
experimental data	2.58	10.16	17.78	27.94	35.56	41.8
FEA result	3.4	9.72	17.34	27.2	35.02	43.03

Fig 8.6.7 Result graph of test sample D8C097-7E6W[ ].

9.4 Weight ratio compression:

Weight ratio of are compared with Standard gantry model with all three cases of model design. The weight difference in case 1 Girder and runway with standard part is 86%,85% and 91% respectively, for case2 and case 3 the difference is 81%,80%,88% and 72%,71%,83% respectively . This study shows that weight and proportionately cost of this parts are also much lower than standard BPG FP model.

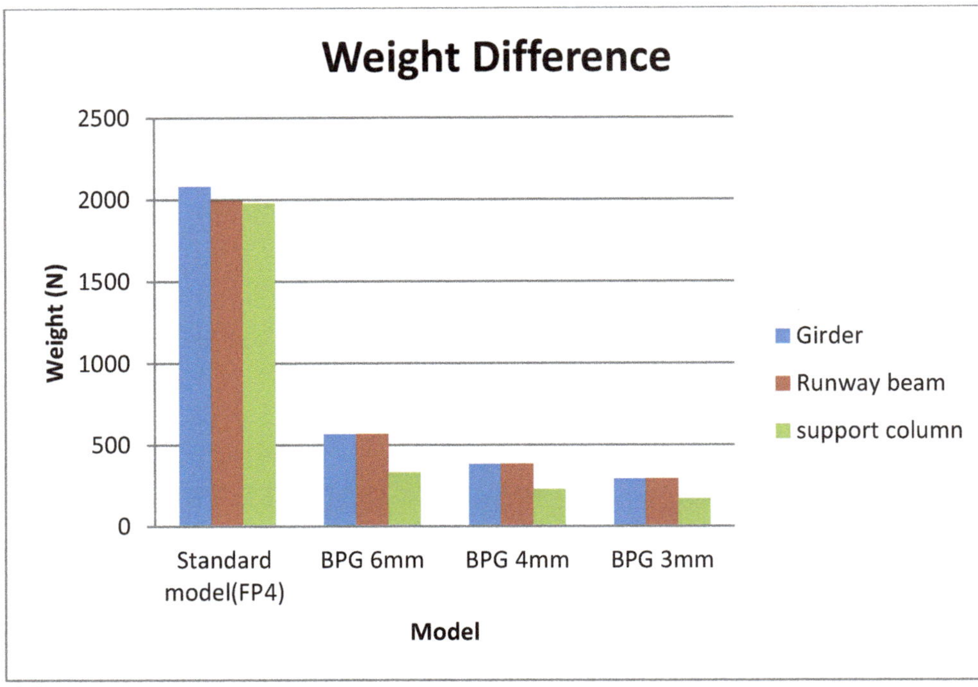

Fig.9.4.1 Weight compression of all BPG models.

10. CONCLUSION:

In this study, Stress and deflection value on the component of Bag Palletizer gantry are computed both by analytical and numerical by FEM method. The analysis of Bag palletizer gantry is done for three cases of varying thickness of C-section. In first case of 3mm thick member the maximum deviation in von mises stress for Girder and Runwaybeam are not exceed than 9.95N/mm^{2}and 8.5N/mm^2 and deviation in deflection are not more than 0.106mm,0.083 respectively. Similarly in second case the maximum devotion of stress and deflection vale for Girder and Runway Beam are 6.74N/mm^2 , 5.62N/mm^2 and0.071mm,0.059mm respectively . Third case shows In all these cases the deflection and stress value effect on section strength. All these three cases compared by weight ratio to standard model are 87 % to 75%.

The finite element model of the testing has been setup and verified by experimental data. The deviation of result is around 4.5%. The FE analysis has been carried out to the cold rolled mild steel beam which has same dimension as BPG required design sample of C-Section. The result obtain from analysis and verification are help for selecting of C-Section dimension.

As the computed stress value are smaller than the allowable stress of the material of Bag Palletizer Gantry component, it is observed that the Bag Palletizer Gantry crane is safe according to FEM norms. This is proves that FEM analysis method is insufficient in some areas. The result obtained from FEM analysis shows that the element type and boundary condition are selected correctly. Considering the data obtain from analysis, the material west can be prevented in newly designed Bag Palletizer Gantry. The Structure is now more reliable and light weight. This is important in means of low cost production and low concept design duration is possible due to use of Statistical Design Institute Triptych tool. This tool decrees half of BPG development time.

11. REFERENCES:

[1] Gabor Jakab, ''*Analysis and design of cold formed C-section member and structure*'', Budapest university of technology and economics ,2009.

[2] B.J.Dias,F.V.De, ''*Design and optimization of a light weight aluminium gantry system*'', Norwegian University of Science and Technology, Oct 2013.

[3] R.A.Kulwiec(ed), ''*Material Handling Handbook*'',2nd Edition,1985.

[4] Cheng Yu, Ben Schafer, ''*Distortional buckling of C and Z member in bending*'', The American Iron and Steel Institute(AISI),Aug2004.

[5] C.Alkin,C.E.Imrak,H.Kocabas,Vol.45, ''*Solid Modeling and Finite Element Analysis of an Overhead Crane Bridge*'' Czech Technical University in Prague,2005.

[6]GUDEL India module catalogue, D,F,E/01.13/NR.0112130

[7]A.Ghiorghe, ''*Optimization design for the structure of an RRR type industrial robot*'',Vol.72,Iss,4.2010.

[8]Power grip design manual, Gates Europe ISO 14001,E2/20099,ED 2005.

[9]AK Steel, ''*Product Data Bulletin*''0130 CR7/07,2007.

[10] J.I.Lerga's ''*Modeling and simulation research on the metal structure of bridge crane*''Technical university ,Sofia,2011.

[11] John R.Hauser and Don Clausing, ''*The House of Quality*'', Harvard Business Review ,May-June 1988.

[12] Gates Rubber Co., Sales Engineering Dept, Rubber Manufacturers Association (RMA), International Organization for Standardization (ISO).

APPENDIX

This section shows design of power drive system use in bag palletizer gantry and its effect on section design.

Power Drive Design:

An early model of bag palletizer gantry uses rack and pinion mechanism for drive system. This mechanism is basically having many disadvantage like heavy weight ,not suitable for powder and food industry and maintenance due to wear and tear is much expensive .also rack and pinion maintenance is time consuming and it will stop all process till maintenance completed. To overcome this disadvantage of rack and pinion by replacing with Belt and pulley drive mechanism. Which has many advantages like light in weight, High Torque Drive (HTD) belt available in market which has high torque caring capacity and this mechanism is easy work freely in any environmental condition. The main difference is about cost of Belt and Pulley drive is less than that of Rack and Pinion drive mechanism. The design and selection of Belt and Pulley is carried out in PowerGrip belt manual.

Design of belt requirement:

Load to carry -2928N

Speed -4m/sec.

Diameter of Pulley -100mm[8]

Length of belt-3714mm

$$L=2C+1.57(2\times D)$$

Where, L= Length of belt

D=Diameter of pulley

C=Center distance

Belt design for above requirement: 85HTD8M-3714-ML [8]

Expected shape of girder for particular Belt and Pulley assembly with hoist assembly arrangement consideration. Few mild steel sections with hoist assembly condition are shown bellow.

Table 11.1.1 Concept section

Section with Carriage plate and pulley arrangement		
Square section	C' Section	I' Section
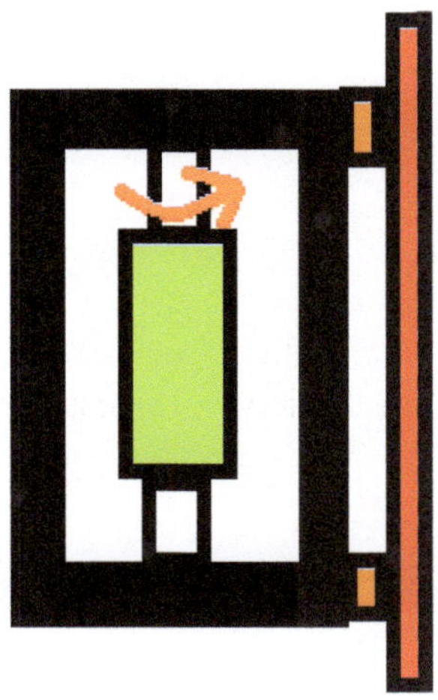	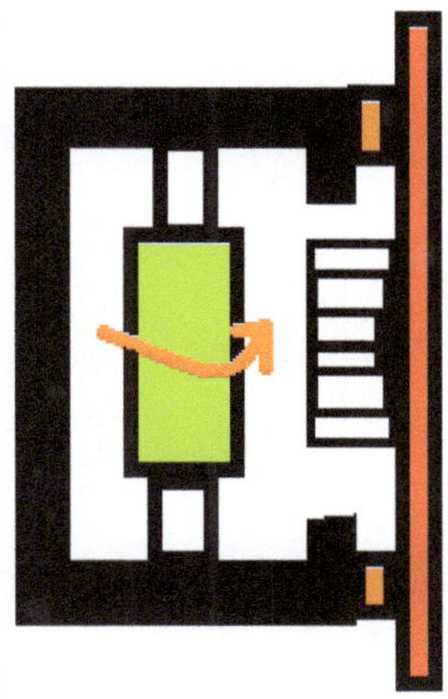	
Belt arrangement is **not Possible** easily.	Belt arrangement is **Possible easily.**	Belt arrangement is **Possible easily** .But not good for guide ways and carriage plate assembly

From above table it is finalized that C-section is preferred for girder profile and hoist assembly point of view. Now, design of C-section dimension for girder profile with consideration of hoist assembly joints, belt and pulley size.

Case 1:

a=200 to 220 mm because, Hoist assembly roller gap is 210 so that gap must be remain constant.

b=100 to 120 mm because, Pulley diameter is 100mm as given.

c=32 to 35 mm because, Guide ways having standard size.

t=3,4,6mm (assume)

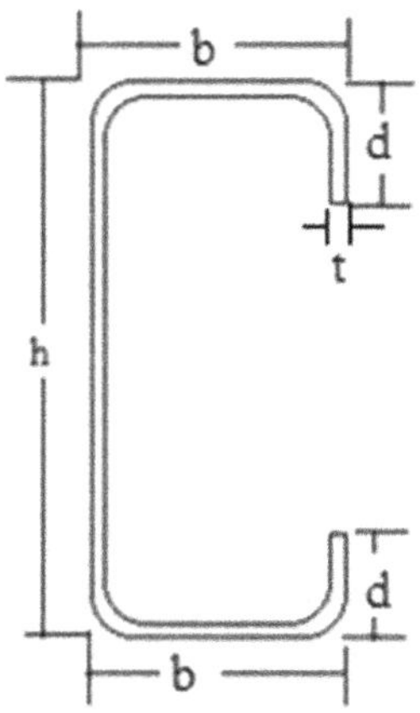

Fig.11.1.1 C-Section Parameter

This above section parameter is design on the basis of power system requirement. But this selected section must be analyzed for actual loading condition. The analysis result get validate by analytical result and if those result are having less than 15% deviation[4] then the designed section and whole Bag Palletizer Gantry is valid for manufacturing.

12

Numerical Investigation of Flow Characteristics over a Square Cylinder with a Detached Flat Plate of Varying Thickness at Critical Gap Distance in the wake at Low Reynolds Number

Nidhul K[1], Sunil A S[2], Kishore V[3]

PG student, Govt. Engineering College Thrissur, nidhul07@gmail.com[1]
Associate Professor, Govt. Engineering College Thrissur, sunilkoni@gmail.com[2]
Assistant Professor, Govt. Engineering College Kannur, kish.venugopal@gmail.com[3]

Abstract

Flow past a square cylinder with and without corner modifications is carried out numerically by using CFD fluent. The wake is generated by a uniform flow of Reynolds number (Re) 150 based on the characteristic length of the cylinder, D. 2D unsteady numerical simulation is done using FVM employing pressure based solver and PISO scheme. A computational grid independence study has been done to obtain a grid resolution which predicts the results without any discrepancies. The flow separation point for the square cylinder with and without corner modifications is obtained. The pressure distribution in the near wake region and around the square cylinder surface is also investigated for different corner geometries. The results are presented in the form of coefficient of pressure Vs Domain length/D, Coefficient of pressure along the cylinder wall and wall shear stress along the cylinder wall. The results indicate that the flow separation in case of square cylinder without sharp corners is delayed. The adverse pressure gradient along the surface of the cylinder and in the near wake region is smaller for round corners. The tangential velocity of square cylinder with sharp corners is large when compared with modified corners.

***Keywords**:* Square Cylinder, Corner modifications, Reynolds number, wake, grid independence, flow separation point, Pressure distribution, Lift coefficient, Drag coefficient.
.

1. Introduction

Vortex shedding behind bluff bodies is of concern for many engineering applications. Bluff bodies are structures with shapes that significantly disturb the flow around them, as opposed to flow around a streamlined body. Fluid flow past a cylindrical object generates vorticity when Re is increased due to the shear present in the boundary layer. This vorticity in the flow field coalesces into regions of concentrated vorticity known as vortices. Further increase of the Reynolds number (Re > 40) makes the steady vortices to become unstable and the flow bifurcates to a time-periodic state, in which opposite-signed vortices are periodically shed from the opposite sides of the near wake; this is known as the primary instability of the wake. The periodic vortex shedding generates oscillatory forces on the cylinder. The forces on the plane of the cross-section of the cylinder can be decomposed into drag– the force acting in the same direction of the free-stream – and lift – the force acting in the cross-stream direction. Due to the existence of these oscillatory forces imposed by the flow, flexible cylinders and rigid cylinders mounted on flexible structures vibrate when immersed in a uniform stream. This vibration occurs predominantly in the cross-stream direction and since the origin of the forces is associated with the vortex shedding, this type of structural response is called vortex-induced vibration (VIV). VIV is a strongly nonlinear phenomenon, since the movement of the cylinder alters the flow field, and the flow field is ultimately responsible for the forces exerted on the cylinder. Vortex shedding is responsible for structural movement of high rise buildings, scour development around bridge piers in channel beds, vibrations of industrial components, acoustic radiation from aircraft landing gear and other related problems. Therefore, it is important to understand and control these flow-induced problems so that engineering design and public comfort can be improved continuously.

As a result, this subject requires investigation, particularly when the near wake is interfered with, e.g., by a flat plate, so that a comprehensive understanding about behaviour of the flow for a wide range of conditions can be obtained. Thus the fluid flow around a cylinder, because of complicated phenomena such as vortex shedding and flow separation behind the cylinder, has been studied by many researchers and scientists. They applied some methods and devices to control this phenomenon. Methods are classified in three groups: (1) passive control, (2) active control and (3) compound control. Passive control techniques do not need any external energy during application. Additional devices in the fluid flow or changing the geometry of the bluff body such as splitter plate, base bleed and roughness are applied in this method. Active control techniques such as EHD actuators and vibrators need external energy to affect the fluid flow. When Active and passive techniques are applied simultaneously, it is called compound method.

Over the last twenty years, a vast amount of studies has been conducted to increase the understanding of different transition processes of the flow past a circular cylinder-experimentally, numerically and theoretically. By contrast, there are very few similar studies found on flow past rectangular cylindrical structures, e.g. the square cylinder, at moderate Reynolds number. Inoue et.al [1] constructed a non-uniform mesh but divided the computational domain into three regions, each with a different grid ratio. At Reynolds numbers less than about 500 there is only one single set of experiments, Okajima et al.[2] reported the influence of Reynolds number on the mean drag coefficient. Mohamed Sukri Mat Ali et.al [3] numerically investigated the sensitivity of the computed flow field to flow parameters for a flow with Reynolds number 150.They constructed computational meshes based on reasonable estimates of cell size and grid stretching ratios. S. Ozono [3] conducted experiments over a circular and rectangular cylinder with a horizontal short thin splitter plate below wake centerline and showed that vortex can be suppressed even when the splitter plate is asymmetrically behind the cylinder. Mittal [4] numerically investigated the effect of varying plate length and its position downstream on the near wake of a circular cylinder at a Reynolds number of 100. Slip boundary condition was imposed on the plate, which may not be relevant to practical applications. Rathakrishnan [5] comprehensively studied the vortex shedding suppression using the splitter-plates attached to the end of the cylinder. Doolan [6] numerically studied flow around square cylinder of side length D with a downstream flat plate of length 0.834D laid at 2.37D from the rear surface of the cylinder. A strong interaction between the shear layers from the square cylinder wall with the flat plate was observed. Mohamed Sukri Mat Ali et.al [7] observed numerically that the critical gap distance Gcr for a detached splitter plate is 2.3D. The plate had no significant effect on the generation of the von Kármán vortex when the separation is beyond 5.6D. Many studies show that it is very important to accurately obtain the value of G_{cr} as flow transits to a new flow regime after critical gap distance. (e.g., Bull et al. [8]; Carmo et al. [9]; Papaioannou et al. [10]; Zdravkovich [11]). The current study aims to investigate the influence of a detached flat plate of length L=D and its varying thickness on the flow characteristics around the square cylinder.

2. Numerical Simulation Procedure

2.1 Flow Field Formulation

The governing equations on the flow field are the continuity and momentum equations (Navier–Stokes equations), which can be written as follows:

$$\frac{\partial \rho}{\partial t} + div(\rho V) = 0 \tag{1}$$

$$\frac{\partial}{\partial t}(\rho u) + div(\rho V u) = -\frac{\partial p}{\partial x} + div(\mu \, grad u) + B_x \tag{2}$$

$$\frac{\partial}{\partial t}(\rho v) + div(\rho V v) = -\frac{\partial p}{\partial y} + div(\mu \, grad v) + B_y \tag{3}$$

Where ρ is the fluid density, μ is the fluid viscosity, V is the velocity vector of the flow field, p is the pressure, and u and v are the velocity components in the x- and y-directions, respectively. B_x and B_y are also the body forces per unit volume, which are negligible in the present study. The fluid is assumed to be incompressible, and its properties has been taken as $\rho=1.225$ kg/m^3 and $\mu=1.7894\times10^{-5}$ kg/m s.

2.2 Limits of the problem

A rectangular domain was used with a length of 38D and a width of 20D, where D=0.04 m is side length of the cylinder. The cylinder center has the coordinates x=10.5D and y=10.5D as shown in fig.1 (from Ref .5). The position of the splitter-plate is specified by G in the horizontal direction, from the trailing edge of the cylinder. The splitter plate length and thickness are D and 0.02D, respectively. Splitter plate thickness (H) is varied in the range $0.02D \leq H \leq 0.08D$.

The fluid flows uniformly (with velocity U_∞) from left to right into the downstream of the domain. Boundary conditions should be enforced at the outlet and the lateral boundaries of the computational domain, as well as on the surfaces of the embedded bodies. At domain inlet and outlet, velocity inlet and pressure outlet respectively have been employed. A symmetry boundary condition at the lateral boundaries of the domain and wall with no-slip condition has been used at the cylinder wall. On the splitter plate, no-slip condition has been applied.

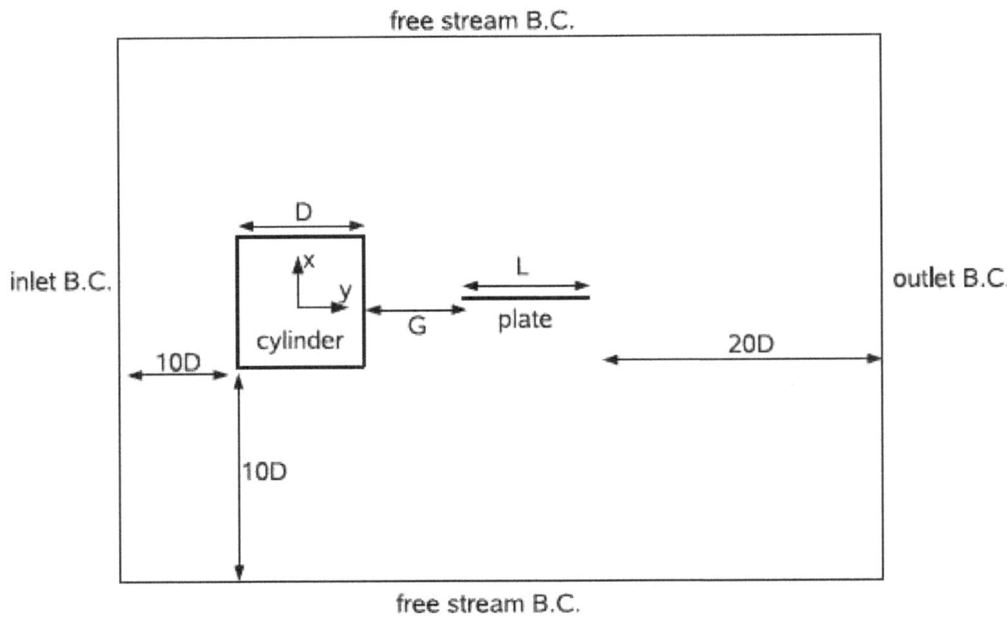

Figure 1: Flow domain for a square cylinder with a splitter plate

2.3 Discretization method

Transforming continuous fluid flow problem into discrete numerical data which are then solved by the computers is known as Discretization. The governing equations have been discretized using the finite-volume method on a fixed Cartesian-staggered grid with non-uniform grid spacing. The grids in the region of the embedded boundaries are sufficiently fine in order to achieve the reasonable accuracy. The spatial discretization has been performed by using multi-block analysis with structured mesh for square cylinder case and for square cylinder with detached flat plate, unstructured mesh near the plate. The temporal discretization has been done in conformity with the second order implicit scheme. Temporal discretization involves the integration of every term in the differential equations over a time step Δt. The simulations were carried out as an unsteady state with a time step (Δt) size of 0.07 sec with 2000 time steps.

2.4 Grid influence study

For predicting the flow field around a square cylinder using numerical analysis, many similar investigations have been made, but the results always show small discrepancies even though the overall global trends are similar. One of the reasons for these discrepancies is the difference in the mesh or grid used for numerical simulation. The mean axial velocity along the center-line of the wake region is shown in fig.2. Six different 2D grids are used ranging from coarse grid to fine grid. The results of the simulations show a certain dependency on the grid resolution for the different grids used. It is observed that in the upstream all the runs predict identical results, whereas a faster recovery of mean axial velocity is predicted when the grid resolution is increased. The mean velocity of recirculation is approximately the same in all runs.

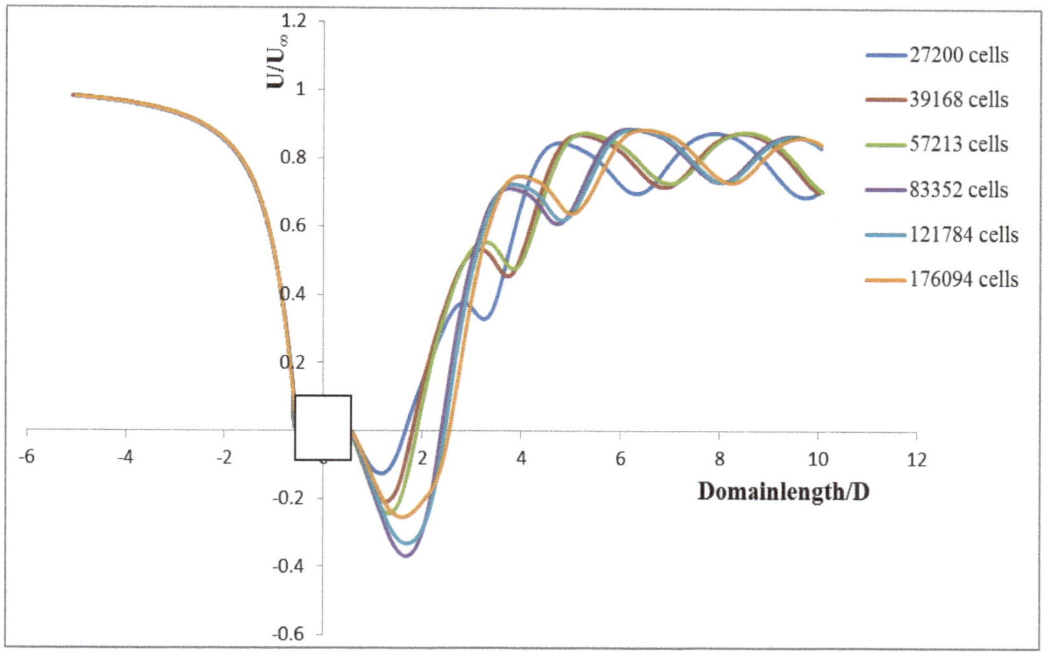

Figure 2: Mean axial velocity distribution along the center-line of the wake

2.5 Verification of the solver

Accuracy of the solver is validated in this section by simulation of flow around a square cylinder. Various parameters such as drag coefficient $Cd= F_d /0.5\rho u^2_\infty D$, pressure coefficient $Cp= P-P_{atm}/0.5\rho u^2_\infty$ and Strouhal number $St=fD/U\infty$ has been used. Grid independency of the solver has been performed and showed that the results are not sensitive to the grid size. Hereinafter, all the results presented have been obtained with fine grid consisting of 83352 cells. Assuming that this grid provides a sufficient grid-independency for the Reynolds number considered in this paper.

The results of the flow around square cylinder were verified using the dimensionless parameters defined above. The value of St obtained by running the simulation at Re=150 was in good agreement with experimental studies of Okajima, 1982and Sohankar et al., 1999. The experimental value of Strouhal number obtained by above mentioned authors was in the range 0.148-0.155 and the present study obtained the value 0.15.Similarly, the value of C_d obtained experimentally by above mentioned authors was 1.40 and the present study obtained the value as 1.58.

3. Results and Discussion

In the first section, the results are demonstrated and analyzed for Reynolds numbers of 40 (non-vortex shedding case) and 150 (vortex shedding case) for square cylinder of side length, D. Second section investigates the effect of downstream flat plate of constant thickness by varying the gap distance (G) along the wake centerline in the range $0.5\ D \leq G \leq 4\ D$ for a constant plate length of L= D. Third section of the paper investigates the effect of varying thickness of the plate placed at critical gap distance from the trailing edge.

3.1 Non-vortex shedding case (Re=40)

Simulation was performed for square cylinder with Re = 40. The principal goal here was to characterize the flow structure around the square cylinder, to obtain the flow variables, pressure variation along the surface of the cylinder and in the near wake region and to obtain drag coefficient. Since a contour plot can quickly reveal regions of high (or low) values of property being studied, fig.3 (a) shows the vorticity magnitude contours for Re=40 which indicate that there is no vortex street. At Re=40, only recirculation eddies are formed immediately downstream of the cylinder as indicated by streamlines shown in fig.4.The U/U_∞ plotted in fig.5 is in agreement with the literature. The fluid comes to a complete stop at the mid-point of incident face of the square cylinder i.e. at the stagnation point. The flow accelerates around the upstream corner ,so much that boundary layer cannot negotiate the sharp corner and separates and reverse flow occurs in the immediate downstream of the cylinder. While moving further downstream of the wake region, flow slowly regains its velocity.

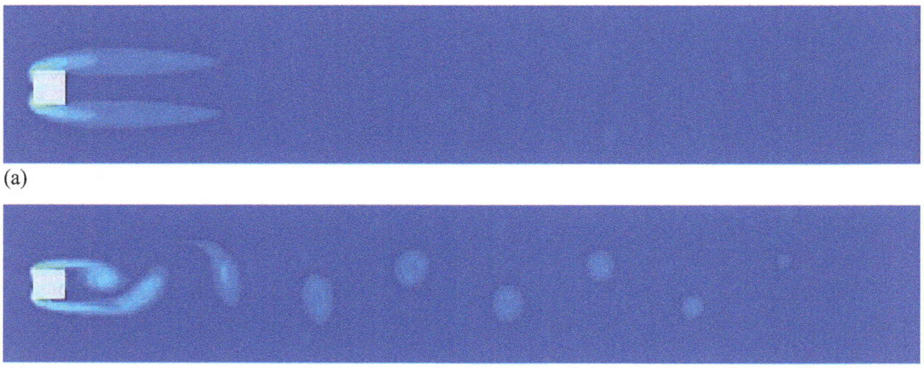

(a)

(b)

Figure 3: Vorticity magnitude contour for (a) Re=40 and (b) Re=150

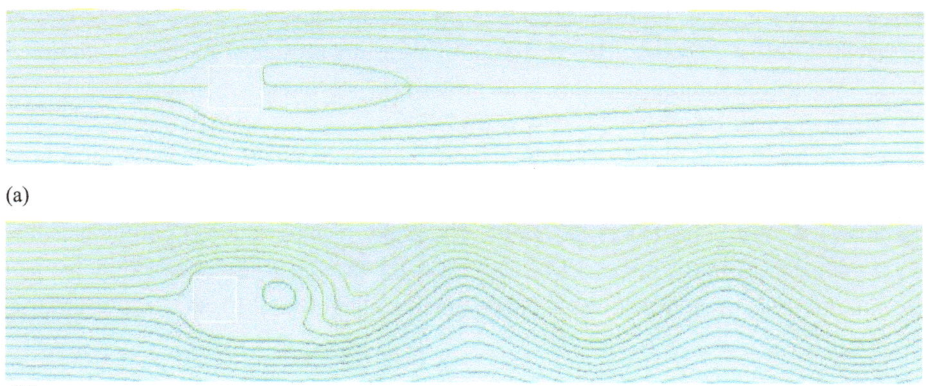

(a)

(b)

Figure 4: Stream function contour for (a) Re=40 and (b) Re=150

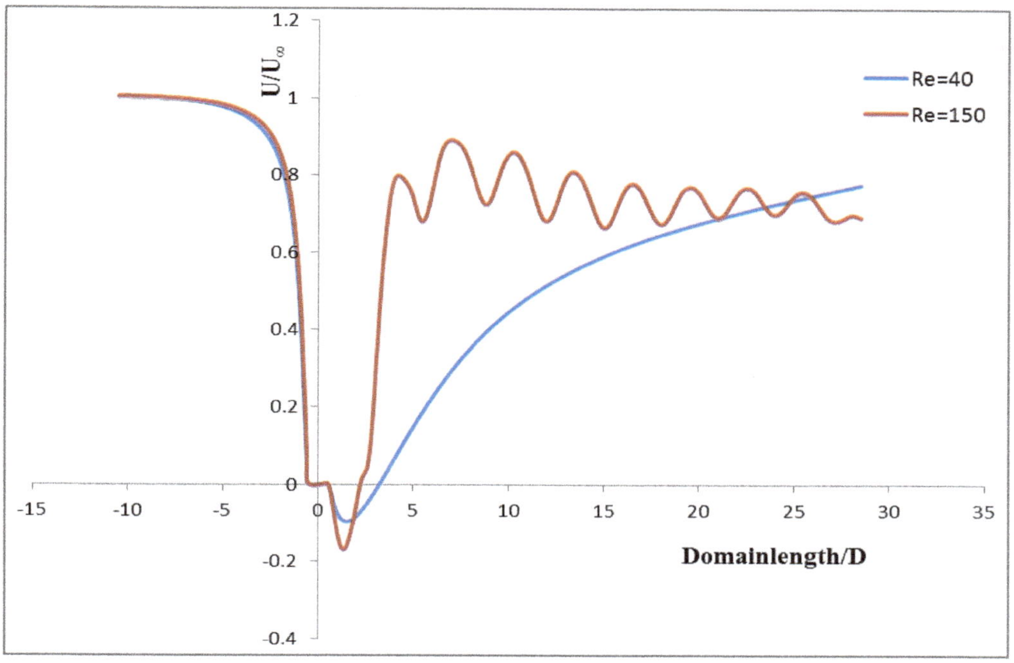

Figure 5: U/U_∞ along the wake center-line for (a) Re=40 and (b) Re=150

Apart from velocity profile, pressure distribution is an important parameter in the study of flow around a cylinder. Pressure changes accordingly with the vortices motion in the vicinity of the body. At stagnation point, located at midpoint of face 0-1as shown in fig.5, flow comes to rest and pressure thus reaches maximum value (By Bernoulli's theorem).It is also noticed that the pressure shows symmetrical distribution; which suggest that problem has been simulated to satisfactory computational time to achieve a convergent solution.

Near the top surface of the cylinder, flow momentum is quite low due to viscous effects and is thus sensitive to changes of pressure gradient. The flow has to move against pressure force in addition to the viscous force. This leads to reduced velocity and wall shear stress. Flow separation occurs when shear stress cannot overcome adverse pressure gradient. At Re=40, flow separation occurs just after top right corner of the square cylinder where wall shear stress is zero as shown in the fig.6. Coefficient of drag (C_d) and coefficient of lift (C_l) for the square cylinder at Re=40 is obtained as 1.76 and 0.0002953.

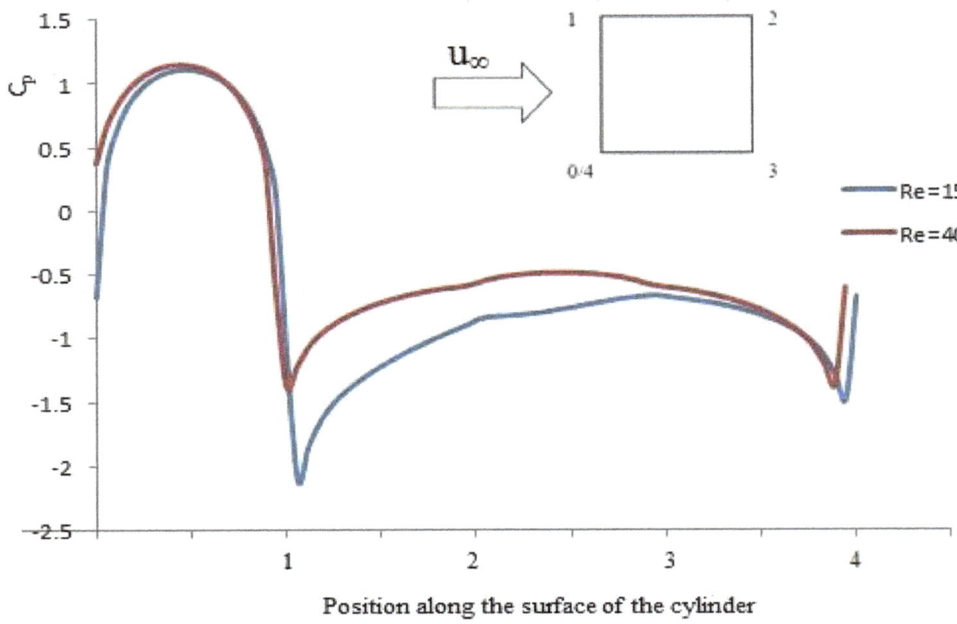

Figure 6: Pressure variation along the surface of the square cylinder for (a) Re=40 and (b) Re=150

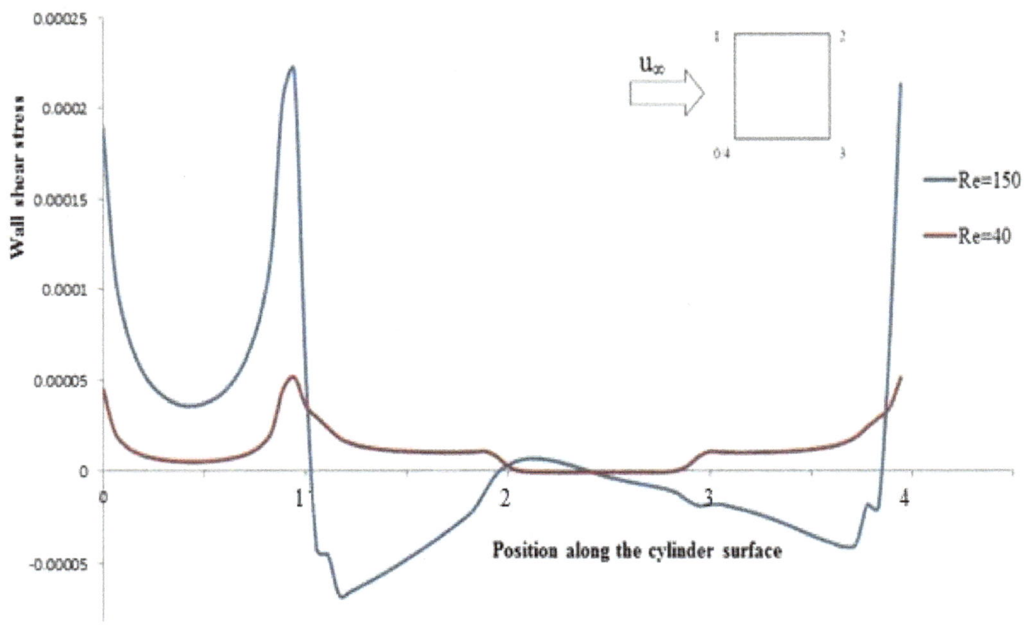

Figure 7: Wall shear stress along the surface of the square cylinder for (a) Re=40 and (b) Re=150

3.2 Vortex shedding case (Re=150)

When Re is increased beyond 40, wake behind the cylinder becomes unstable. Wake develops a slow oscillation in which the velocity is periodic in time and downstream distance, with amplitude of oscillation increasing downstream. Oscillating wake rolls up into two staggered rows of vortices with opposite sense of rotation. Because of similarity of the wake with the footprints in a street, staggered row of vortices behind the square cylinder is called as Von-Karmann Vortex Street. Eddies periodically break off alternately from 2 sides of the cylinder as shown in fig.3. While an eddy on one side is shed, that on the other side forms resulting in an unsteady flow near the cylinder. As vortices of opposite circulations are shed off alternately from 2 sides of the cylinder, results in an oscillating "lift" or lateral force which can be obtained by plotting variation of C_l Vs flow simulation time. Fig.4 shows the streamline pattern for Re=150 indicating a rotating lump of fluid shed from the top portion of the cylinder. From fig.5 it is clear that velocity in the wake region of cylinder at Re=150 is lower than that of Re=40.It also indicates that vortices are shed over a much larger length of the wake.

The pressure variation near the top left corner and top surface of the cylinder is much larger creating a greater adverse pressure gradient (than that of Re=40) opposing the flow as shown in fig.6.Hence the flow around the square cylinder at Re=150 separates off much earlier i.e. at the top left corner of the cylinder as wall shear stress approaches zero after point 1 along cylinder surface as shown in fig.7.The values of C_d, C_l and Strouhal number obtained by simulation are compared with experimental and simulation work of other authors.

Table 1: Comparison of the Present study with other previous studies and experiments

Study	St	C_D	C_L
Experiments(Okajima, 1982; Sohankar et al., 1999)	0.148-0.155	1.40	-
Sohankar et al. (1998)	0.165	1.44	0.230
Doolan (2009)	0.156	1.44	0.296
Inoue et al. (2006)	0.151	1.40	0.40
Ali et al. (2009)	0.160	1.47	0.285
Present study	0.150	1.58	0.196

3.3 Effect of detached flat plate on vortex shedding

Simulation was performed for a square cylinder at Re=150 with a detached flat plate of constant thickness along wake center-line at varying gap distance $0.5D \leq G \leq 4D$.The vorticity magnitude contours for each gap distance was obtained as shown in fig.8.

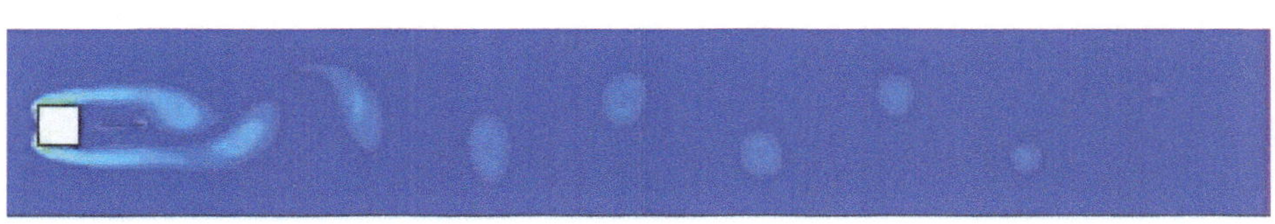

(a)

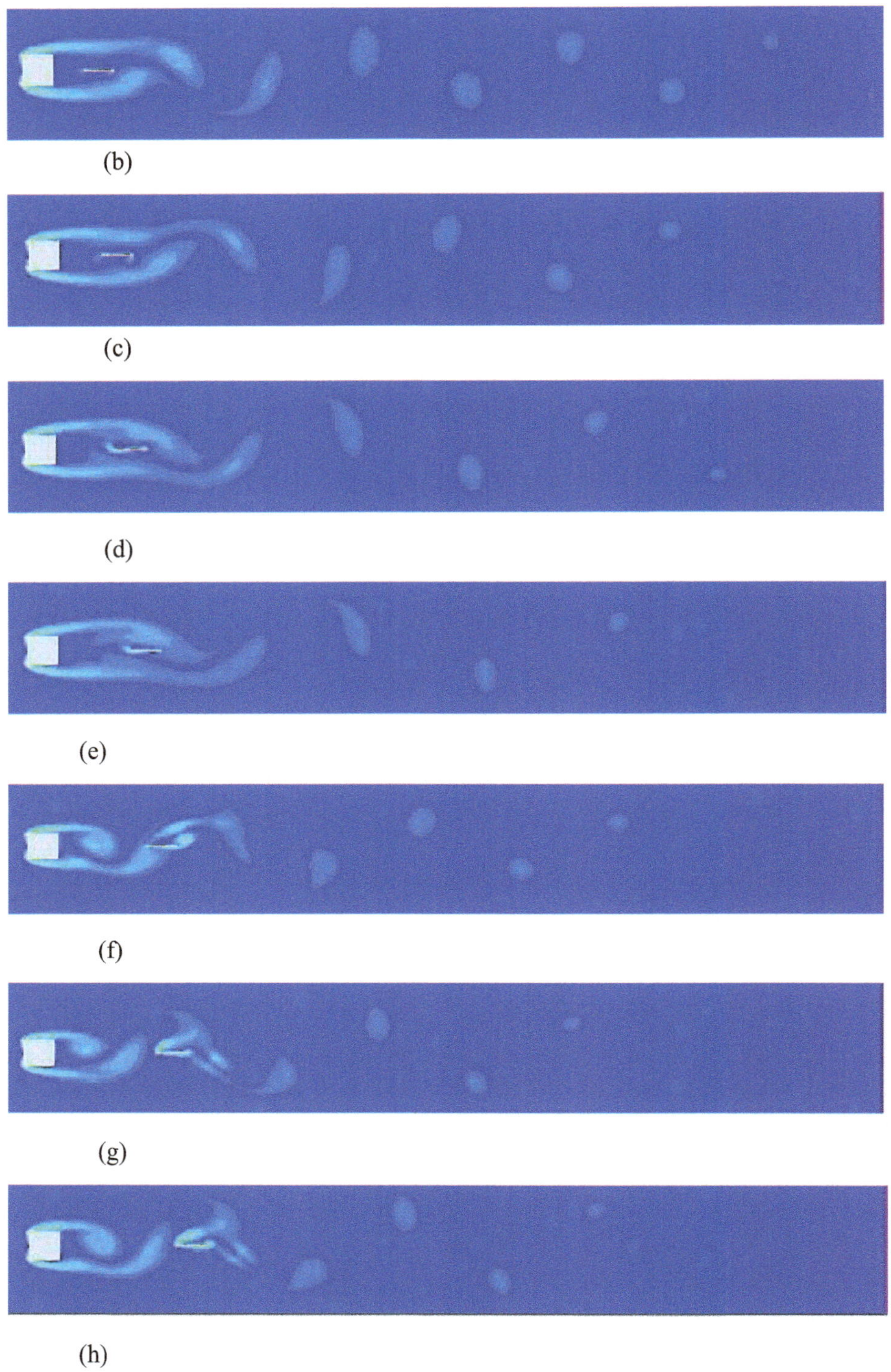

Figure 8: Vorticity magnitude contours of square cylinder with detached plate at varying gap distances (a) G=0.5D (b) G=1D (c) G=1.5D (d) G=2D (e) G=2.5D (f) G=3D (g) G=3.5D (h) G=4D

The space between cylinder and the plate provides an opportunity growing vortex to entrain fluid from other side of the cylinder through the gap and thus there is a change in vortex shedding pattern from that of square cylinder without a plate in the wake region. For close-proximity gap distances, G ≤ 1.5D, there is no direct fluid entrainment into vortex core, as it is prevented by the plate. Beyond gap distance of 2.5D, fluid entrainment process via the gap generates additional vortices in the vicinity and a secondary vortex accumulates at the plate leading edge. The variation of Strouhal number with gap distance is shown in fig.9.

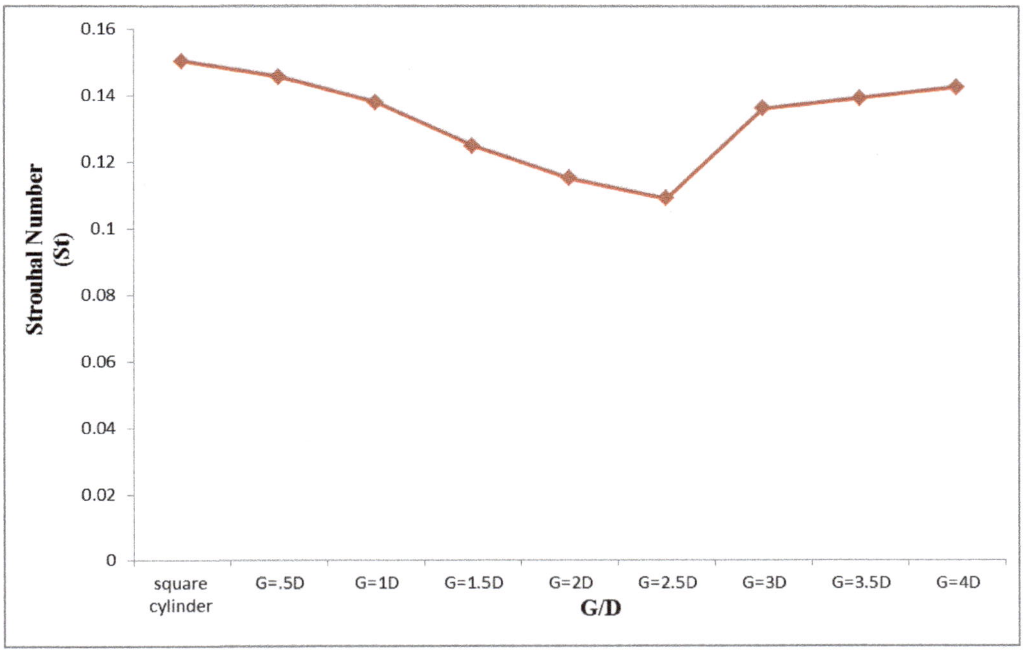

Figure 9: Strouhal number variation for varying gap distances

For 0.5 ≤ G ≤ 2.5D Strouhal number decreases with gap distance. As plate moves closer to core of the growing vortex i.e. as G increases interaction between shear layers is reduced and shear layers interact with the plate. These effects delay vortex formation process that consequently reduces Strouhal number upto G=2.5D.Beyond G=2.5D there is sharp increase in St. This is due to blockage effect created by the plate which reduces convective speed of the shear layers. However the effect slowly weakens as gap distance further increases. The sharp increase in Strouhal number at G=3D is also due to secondary vortex accumulating at the leading edge of the plate. The variation of coefficient of drag with gap distance is shown in fig. 10 and fig.11 respectively.

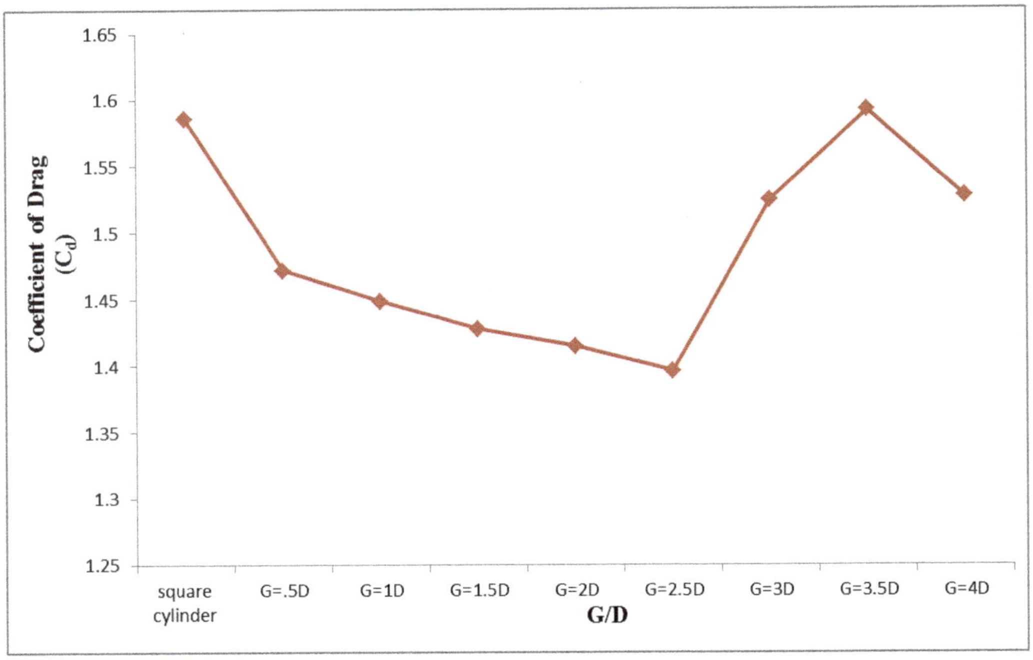

Figure 10: Coefficient of drag variation for varying gap distances

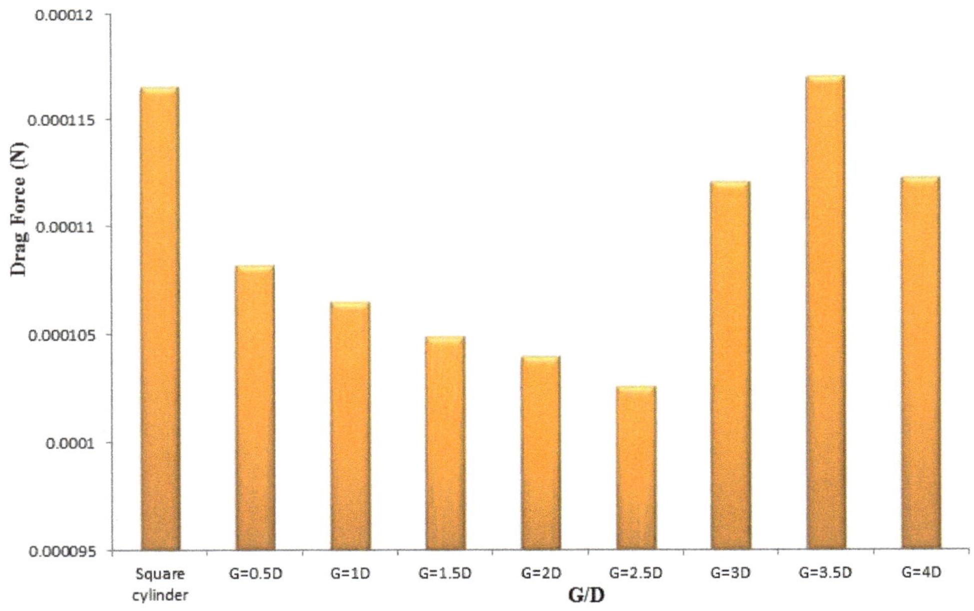

Figure 11: Drag Force variation for varying gap distances

3.4 Effect of thickness of detached flat plate on vortex shedding

The last part of this paper investigates the effect of varying the splitter plate thickness in the range $0.02D \leq H \leq 0.08D$ placed at critical gap distance. The variation of Cp in the near wake region for different plate thickness is shown in fig.13.It is evident that there is no significant variation of pressure in the near wake region when the plate thickness is varied. There is only 1.5% increase of drag coefficient when the plate thickness is increased to H=0.04D.Beyond H=0.04D there is no change in the value of drag coefficient. Similarly there is 4.16 % increase in the value of Strouhal number when plate thickness is increased to H=0.04D. Beyond H=0.04D there is no change in the value of St.

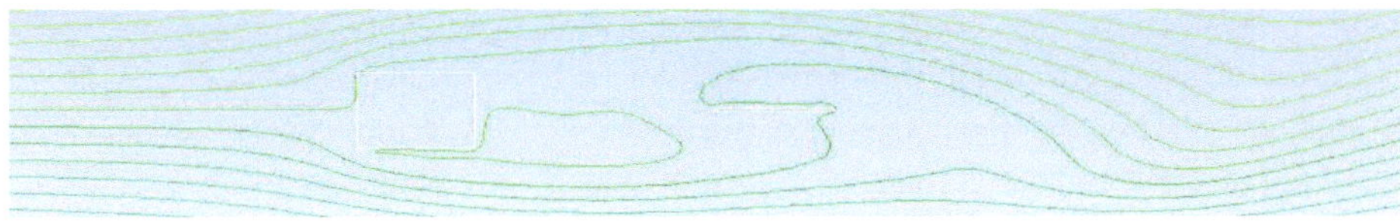

(a)

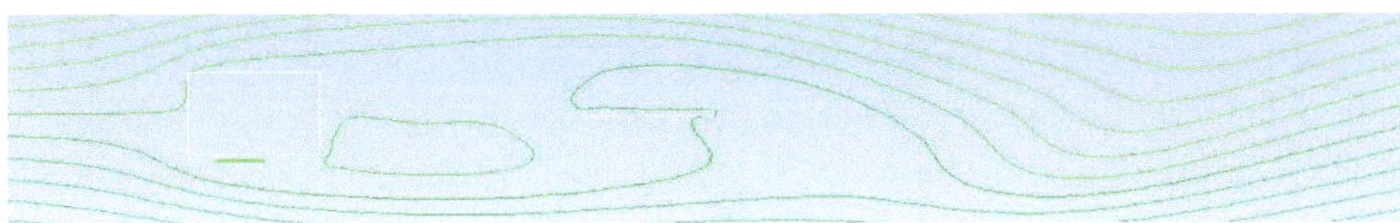

(b)

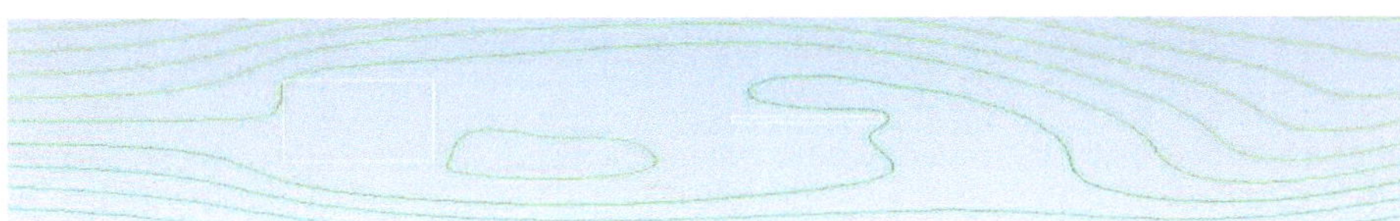

(c)

Figure 12: Stream function contour for (a) H=0.04D (b) H=0.06D and (c) H=0.08D

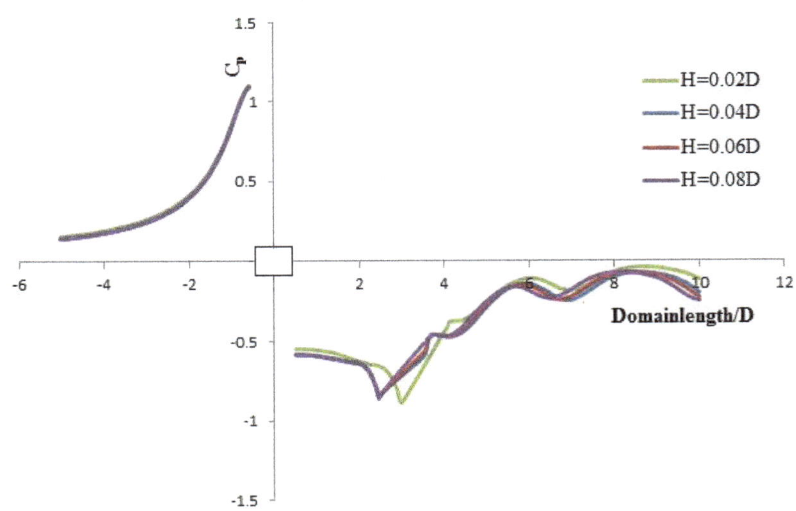

Figure 13: Variation of C_p in the near wake region for varying thickness of the plate

4. Conclusion

The major conclusions are drawn as follows:

(1) At Re=40, only recirculation eddies are formed immediately downstream of the Square cylinder and there is no periodic vortex shedding. For Re>40, flow around square cylinder becomes unstable and results in an oscillating flow with amplitude of oscillation increase downstream. This exerts a force on the cylinder in the lateral direction.

(2) The flow separates at top right corner of square cylinder i.e. flow wraps the cylinder at the top and bottom face at Re=40.But at Re=150, the flow separates much earlier i.e. at top left corner since the wall shear approaches zero after point 1 on the cylinder. Since it is due to this flow separation, rotating lump of fluid masses are formed, flow separation on a square cylinder can be delayed by making the corners smooth.

(3) Detached flat plate suppresses the vortex shedding and reduces the flow induced forces by interrupting the regular vortex shedding. The flow induced force and Strouhal number for the flow around square are cylinder greatly dependent on the position of the flat plate. As gap distance increases, the drag force and Strouhal number decreases and reaches a minimum value at G=2.5D. This gap distance is called the Critical gap distance (G_{cr}) since beyond this gap distance, both Strouhal number & F_d increases. The steep increase in St beyond G_{cr} is due to accumulation of a secondary vortex on the leading edge of the flat plate.

(4) A reduction of 12 % and 27.34 % respectively for drag force and St was obtained by embedding flat plate at G_{cr}=2.5D downstream of the cylinder.

(5)There is no significant variation in flow pattern and flow induced forces when the plate thickness was varied at critical gap distance.

References

[1] Inoue, Iwakam, and Hatakeyama (2006). "*Aeolian tones radiated from flow past two square cylinders in a side-by-side arrangement*" Physics of Fluids, *18(4), 046104.*

[2] Okajima, A., 1982, *Strouhal numbers of rectangular cylinders*. Journal of Fluid Mechanics *123, 379–398.*

[3] Ozono, S., 1999. *Flow control of vortex shedding by a short splitter plate asymmetrically arranged downstream of a cylinder*. Physics of Fluids *11, 2928–2934.*

[4] Mittal, S., 2003. *Effect of a ''slip'' splitter plate on vortex shedding from a cylinder*. Physics of Fluids *15, 817–820.*

[5] Rathakrishnan, E., 1999, "*Effect of Splitter-Plate on Bluff body Drag*" AIAA Journal., *37(9), pp. 1125–1126.*

[6] Doolan, C.J., 2009. *Flat-plate interaction with the near wake of a square cylinder*. AIAA Journal *47, 475–478.*

[7] Mohamed Sukri Mat Ali , Con J. Doolan & Vincent Wheately,2012, "*Low Reynolds number flow over a square cylinder with a detached flat plate*", International Journal of Heat and Fluid Flow, International Journal of Heat and Fluid Flow *36 (2012) 133–141.*

[8] Bull, M.K., Blazewicz, A.M., Pickles, J.M., Bies, D.A., 1996. *Interaction between a vortex wake and an immersed rectangular plate.* Experimental Thermal and Fluid Science *12, 209–220.*

[9] Carmo, B.S., Meneghini, J.R., Sherwin, S.J., 2010. *Secondary instabilities in the flow around two circular cylinders in tandem.* Journal of Fluid Mechanics *644, 395–431.*

[10] Papaioannou, G.V., Yue, D.K.P., Triantafyllou, M.S., Karniadakis, G.E., 2006. *Three dimensionality effects in flow around two tandem cylinders.* Journal of Fluid Mechanics *558, 387–413.*

[11] Farhadi, M., Sedighi, K., Fattahi, E., 2010. *Effect of a splitter plate on flow over a semi-circular cylinder* Proceedings of the Institution of Mechanical Engineers, Part G: Journal of Aerospace Engineering *224, 321 330.*

[12] Zdravkovich, M.M., 1987. *The effects of interference between circular cylinders in cross flow.* Journal of Fluids and Structures *1, 239–261.*

[13] Behzad Ghadiri Dehkordi and Hamed Houri Jafari 2010, *On the Suppression of Vortex Shedding From Circular Cylinders Using Detached Short Splitter-Plates.* Journal of fluids Engineering, *April (2010) Vol. 132 / 044501-1.*

[14] Sohankar, A., Norberg, C., Davidson, L., 1999. *Simulation of three-dimensional flow around a square cylinder at moderate Reynolds numbers.* Physics of Fluids *11, 288–306.*

SPEED CONTROL OF A SEPARATELY-EXCITED DC MOTOR

Monaaf D.A. Al-Falahi[1]

[1]*Faculty of Engineering, Multimedia University, Malaysia, m.o.n_1monaf@hotmail.co.uk*

Abstract

This paper proposes two methods of speed controlling for separately excited dc motor. The methods are provided with their schematics and block diagram, the construction of each scheme was made in breadboard at the first. This eases the testing and troubleshooting of the circuit, and each scheme was constructed in separate breadboard. The results was obtained by using the oscilloscope and the digital multimeter, after assuring of the results obtained, the next step was to fabricate the PCB and demonstrate it in separate boards and jumper wires were used to route the connectors. It was found that the results obtained after fabricating the PCB is similar to that while testing on the breadboard. Afterwards testing the motor was done for the both schemes. The two methods are contributed to the same major factor on controlling the speed of the dc motor, and it was achieved by varying the firing angle, and hence triggers the SCR. Regardless the construction of each scheme, the results obtained proves the theoretical part, and scheme 2 is considered cost effective than scheme 1 and on the other hand it is more reliable in term of size, construction and components availability and accessibility.

Keywords: Thyristor, Firing angle, Half wave controlled rectifier.

1.0 Introduction

An electric motor is an electromechanical device that converts electrical energy to mechanical energy. Electric motors are mainly divided into two types namely, DC motors and AC motors. However, the general working mechanism is the same for all motors. Speed control of an electric motor is very important in many systems and applications such as fans and washing machines. Systems used for motion control are called drives. Hence, drives that use electric motors to control motion are called electric derive [1]. Because they are easy to control, DC motors are widely used in applications where the speed of the motor is to be controlled. DC motors are, in turn, divided into several sub-categories, namely, series motors, long-shunt compound motors, short-shunt compound motors and separately excited DC motors. Two methods of speed control of separately excited DC motors are applicable when the separately excited DC motor is supplied by a constant voltage, namely, armature control method and field control method [2].

In armature control method, a resistor can be installed in series with the armature to vary the voltage supplied to the armature. Using this method we can have a speed that is less than the nominal speed of the motor. However, lot of power lost in the series resistor makes this method inefficient.

In field control method, the idea is in controlling the supplied voltage to the field winding. The main concept used in this method is the inverse proportionality between the motor's speed and the amount of flux. When the supply voltage is small, amount of flux is small as well, and hence the motor's speed is high [2]. Using this method, a speed higher than motor's nominal speed can be achieved.

Some important applications of DC motors speed control are rolling mills, paper mills, mine winders, hoists, machine tools, traction, printing presses, textile mills, excavators, and cranes. Fractional horsepower dc motors are widely used as servo motors for positioning and tracking [3] [4].

2.0 A separately-excited DC motor circuit components and its operation

2.1 Controlled rectifier

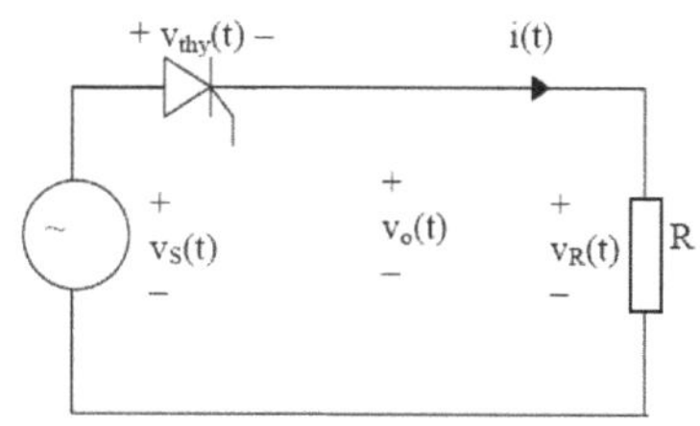

Figure 1: Half-wave controlled rectifier

A single phase half controlled rectifier as shown in Figure 1 above is used in our paper. Thyristor is a controlled rectifier in the circuit. It has two main terminals which is anode, cathode and a controlling gate terminal. A thyristor can only forward bias; therefore the current can only flow in the forward direction, which is from anode to cathode [4]. A thyristor will conduct in the forward direction when a gate current is applied to the gate terminal. However, it will block the forward current when a smaller current pulse enters the gate terminal. Therefore, the gate terminal plays an important role to switch on or off the thyristor. With that, a thyristor is different compared to a diode, and works as a controlled rectifier [3].

Furthermore, when we increase the gate pulse, it will reach a "latching" level. At this state, even if we remove the gate pulse, but the thyristor will still remain on. Once established, the anode-cathode current cannot be interrupted by any gate signal. We can only switch off the thyristor with an external switch. In other words, the non-conducting state can only be restored after the current flowing through the thyristor is reduced to zero. For example, to switch off the thyristor, we must switch off the power supply. Moreover, a thyristor acts as a close switch when conducting. It only has a forward voltage drop of 1V or 2V over wide range or current. Despite the low volt drop in the "on" state, heat is dissipated. Therefore an efficient heat sink is required, to cool down and protect the thyristor.

Power transistors such as power BJT can control the current flow through by adjusting the base current; therefore it is better than thyristors. However, we use thyristor instead of power transistor because a few reasons. Among the reasons is because the thyristor is **lower cost compare to power BJT**. On the other hand, the thyristor has the **higher power capability** compare to BJT. Thyristors allows a very larger current and power to flow through. Furthermore, thyristors have a **low triggering power**. In other words, a thyristor needs a low gate pulse to switch on. Besides that, a thyristor also have a **low on-state voltage drop [5]**.

2.1.1 Single Phase Half-Wave Controlled Rectifier

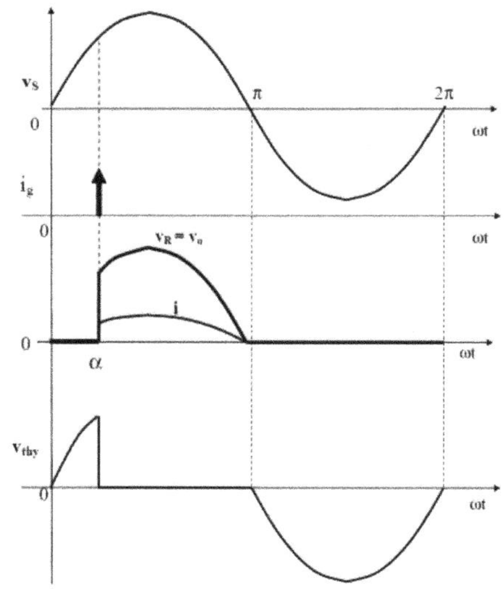

Figure 2: Waveforms for the single-phase half wave circuit

In figure 3, the first waveform shows the voltage input, the second waveform shows the output current from the thyristor and the 3rd waveform shows the load voltage of the circuit. We can observe that there are only the positive half-cycles of the AC supply voltage to the load. However, it is not smooth, but is DC in the sense that it has a positive mean value. By changing the firing angle (α), which measured from the zero crossing of the supply voltage, the mean voltage can be controlled. The circuit is Figure II is called "single-pulse" or half-wave circuit because of the output waveform shown in the Figure III gives only one peak in the rectified output for each complete cycle of the source [6] .

2.1.2 Calculation:

a. Half wave controlled rectifier:

$$V_R = \frac{1}{2\pi}\int_0^{\pi} Vm \sin\theta\delta\theta$$
$$= \frac{1}{2\pi} Vm[-\cos\theta]_0^{\pi}$$
$$= \frac{2Vm}{2\pi}$$
$$= \frac{Vm}{\pi} \quad (1)$$

Since $V_m = \sqrt{2}Vrms$,

Therefore,

$$V_R = \frac{\sqrt{2}}{\pi} Vrms \quad (2)$$

b. Half wave controlled rectifier with firing angle α:

$$\begin{aligned} V_R &= \frac{1}{2\pi}\int_{\alpha}^{\pi} Vm \sin\theta\delta\theta \\ &= \frac{1}{2\pi} Vm[-\cos\theta]_{\alpha}^{\pi} \\ &= \frac{Vm}{2\pi}(1+\cos\alpha) \\ &= \frac{Vrms}{\sqrt{2}\pi}(1+\cos\alpha) \end{aligned} \quad (3)$$

2.2 Basic structure and operation of thyristor

Thyristors are usually three-terminal devices that have four layers of alternating p-type and n-type material (i.e. three p–n junctions) comprising its main power handling section. The thyristor circuit symbol is shown in figure 3(a). Generally, the operation of thyristors is as follows; when a positive voltage is applied to the anode (with respect to cathode), the thyristor is in its forward-blocking state. The center junction, J2 (see Figure 3(b)) is reverse biased. In this operating mode the gate current is held to zero (open circuit). In practice, the gate electrode is biased to a small negative voltage (with respect to the cathode) to reverse bias the GK-junction J3 and prevent charge-carriers from being injected into the p-base. In this condition only thermally generated leakage current flows through the device and can often be approximated as zero in value (the actual value of the leakage current is typically many orders of magnitude lower than the conducted current in the on-state). As long as the forward applied voltage does not exceed the value necessary to cause excessive carrier multiplication in the depletion region around J2 (avalanche breakdown), the thyristor remains in an off-state (forward-blocking). If the applied voltage exceeds the maximum forward-blocking voltage of the thyristor, it will switch to its on-state. However, this mode of turn-on cause non-uniformity in the current flow, is generally destructive, and should be avoided. When a positive gate current is injected into the device, J3 becomes forward biased and electrons are injected from the n-emitter into the p-base. Some of these electrons diffuse across the p-base and get collected in the n-base. This collected charge causes a change in the bias condition of J1. The change in bias of J1 causes holes to be injected from the p-emitter into the n-base. These holes diffuse across the n-base and are collected in the p-base. The addition of these collected holes in the p-base acts the same as gate current [7]. The entire process is regenerative and will cause the increase in charge carriers until J2 also becomes forward biased and the thyristor is latched in its on-state (forward-conduction) [5] [6]. The regenerative action will take place as long as the gate current is applied in sufficient amount and for a sufficient length of time. This mode of turn on is considered to be the desired one as it is controlled by the gate signal.

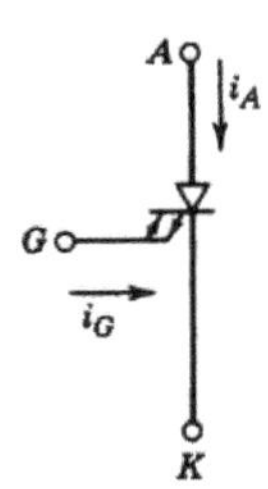

(a) Thyristor symbol

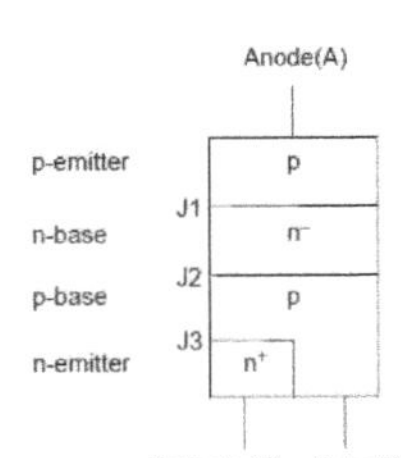

(b)Simple cross section of a typical thyristor

Figure 3: Circuit symbol and cross section of thyristor

The uniqueness of the thyristor lies principally in its Volt-Amp characteristic, which shown in Figure 4. In the reverse direction the thyristor appears similar to a reverse- biased diode, which conducts very little current until avalanche breakdown occurs. Besides that, SCR can exhibit very high internal impedance, with perhaps a slight reverse blocking current. However, if the reverse breakdown voltage is exceeded, the reverse current rapidly increases to a large value and may destroy the SCR. In the forward bias condition (gate is grounded), the internal impedance of the SCR is very high with a small current flowing called the forward blocking current. When the forward voltage ($+V_F$) is increased beyond the forward break over voltage point, an avalanche breakdown occurs and the current from the cathode to anode increases rapidly. A regenerative action occurs with the conduction of p-n junctions and the internal impedance of the SCR decreases. If the gate current is allowed to flow, the forward break over point will be smaller. The larger the gate current flows, the lower the point at which forward break over will occur as illustrated in Figure 4 bellow. Normally, SCRs operated with applied voltage lower than the forward break over voltage point (with no gate current flowing) and the gate triggering current is made sufficiently large to ensure complete turn on. Figure 4 bellow illustrates the volt-ampere characteristics curve of an SCR. The vertical axis (+I) represents the device current, and the horizontal axis (+V) is the voltage applied across the device anode to cathode. The parameter I_F defines the RMS forward current that the SCR can carry in the ON state, while V_R defines the amount of voltage the unit can block in the OFF state [7] [8].

The holding current I_H, for more details, represents the minimum current that can flow through the thyristor and still maintain the device in the ON state. This current value and the accompanying voltage across the device, termed the holding voltage V_H' represent the lowest possible extension of the ON-state portion of the Volt-Amp characteristic.

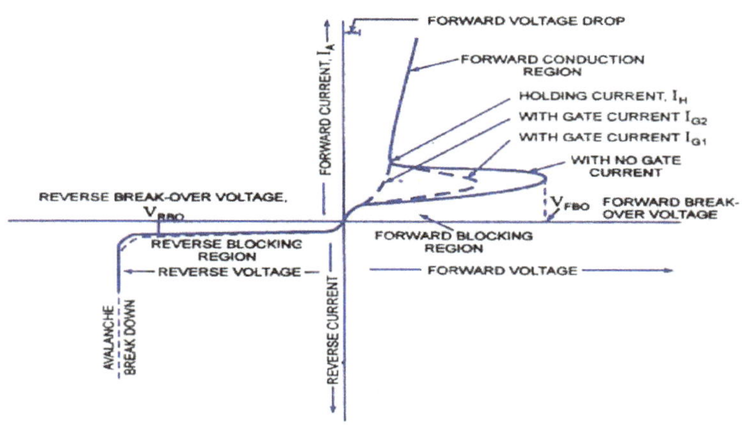

V-I Characteristics of SCR

Figure 4: Volt-Amp Characteristics of SCR

2.3 Phase Control

In SCR Phase Control, the firing angle which the SCR is triggered, determines the amount of current which flows through the device during the half-cycle. It acts as a high-speed switch which is open for the first part of the cycle, and then closes to allow power flow after the trigger pulse is applied. Figure 2.5.1, shows the AC voltage waveform being applied with a gating pulse at 45 degrees. There are 360 electrical degrees in a cycle; 180 degrees in a half-cycle. The number of degrees from the beginning of the cycle until the SCR is gated ON is referred to as the firing angle, and the number of degrees that the SCR remains conducting is known as the conduction angle. The earlier in the cycle the SCR is gated ON, the greater will be the voltage applied to the load. Figure 5 shows a comparison between the average output voltages for an SCR being gated on at 30 degrees as compared with one which has a firing angle of 90 degrees. Note that the earlier the SCR is fired, the higher the output voltage applied to the load. For accurate SCR gating, the Firing Circuit must be synchronized with the AC line voltage being applied anode to- cathode across the device. Without synchronization, the SCR firing would be random in nature and the system response variable. In closed-loop systems, such as motor control, an Error Detector Circuit or/ Zero crossing detector Unit computes the required firing angle based on the system set-point and the actual system output [6] [9]. The firing circuit is able to sense the start of the cycle, and, based on an input from the Error Detector, delaying the firing pulse until proper time in the cycle will provide desired output voltage.

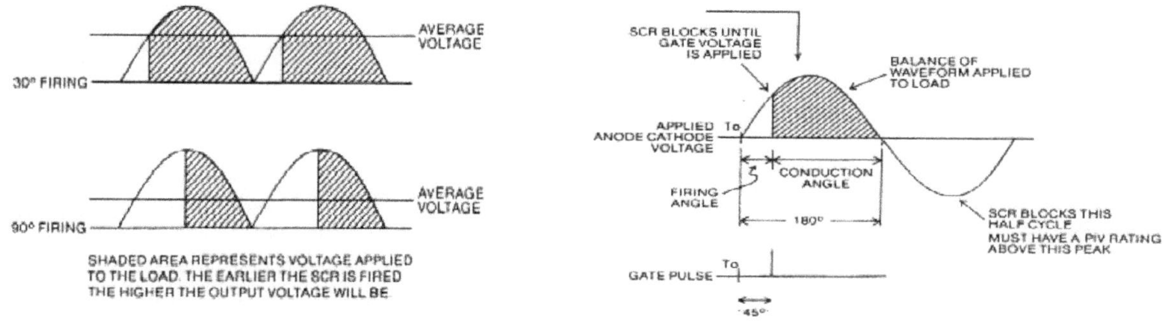

a. Voltage waveform at 45 degrees

b. Voltage waveform at 30 and 90 degrees

Figure 5: A comparison between the average output voltages for an SCR being gated on at 30 and 90 degrees firing angle.

3.0 Details of the design

3.1 Circuit construction

For the firing angle control, information of zero crossing point of an AC input is vital. This project requires student to design circuit for each blocks as shown in figure 6 (i.e. Zero crossing detector, firing angle circuit and pulse shaping). The firing angle can be varied from 0 to 180 degree through a potentiometer. The block diagram of the overall drives system to be developed under this project is shown in figure 6 scheme 1.

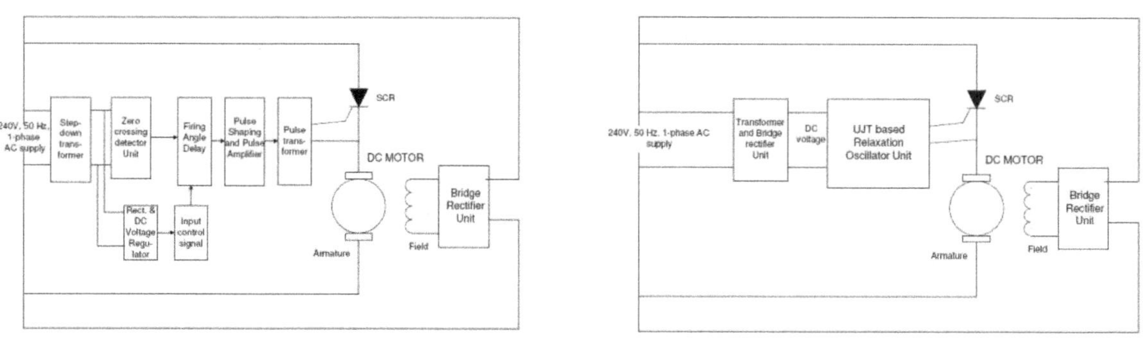

Scheme 1

Scheme 2

Figure 6: The two schemes for speed control of DC separately excited motor

UJT based Relaxation Oscillator can be used for providing triggering pulses to the gate of the SCR as shown in figure V above. Depending upon the delay angle, the speed of the DC motor varies. We then design and construct the complete circuit including dc power supplies as shown if figure 6 scheme 2 above. Next we will go in detail to explain each block in block diagrams above.

3.2 Transformer and bridge rectifier unit

This section is consisting of the step down transformer, half wave rectifier circuit and the DC voltage regulator unit. The step down transformer rating as, primary of: 230V and the Secondary: 12V-0-12V (AC), the secondary winding of the transformer is connected to rectifier unit, A transformer rectifier unit (TRU) is used to convert AC into relatively smooth DC. This device takes the mains 240 V-AC and converts it to approximately 12 V-DC to supply the main circuit. This is achieved by a transformer, which first steps down the AC voltage to a reasonable level. Then converts it via a bridge rectifier and regulate it via the voltage regulator LM7812 assembly into DC. Diodes in their basic forms are used for rectification (or conversion) of AC into DC, the diodes offer an easy path for current to flow in one direction and offer a high resistance path in the opposite direction. The (LM7812) Linear IC Voltage Regulator, used to maintain the output voltage from the dc supply over a wide range of load and line variations. One of the popular positive fixed three-terminal voltage regulators is known from the first two digits **78**XX series and the other two digits XX**12** represent the regulated output voltage [10]. Figure 7 represents the DC voltage regulator schematic circuit.

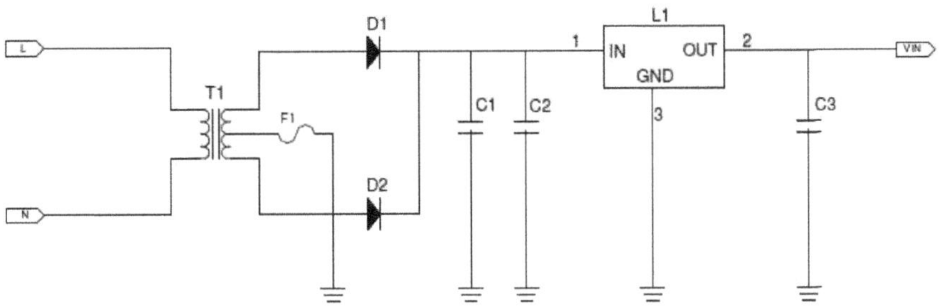

Figure 7: The DC voltage regulator schematic circuit

The two capacitors C1 and C2 in (Figure 7) act as a filters, C1 is used to smooth the dc voltage and remove the unwanted lower frequency noise to the unregulated power supply while C2 is used to provide rejection of transients in the input of the unregulated supply. Capacitor C3 is commonly used to provide improved output impedance, rejection of transients and shorting any very high frequency to ground and preventing it from getting back into the power supply line and travelling to other circuits. For simple explain on the system of the bridge rectification unit, Figure 8 shows a simple block diagram of a regulated power supply system.

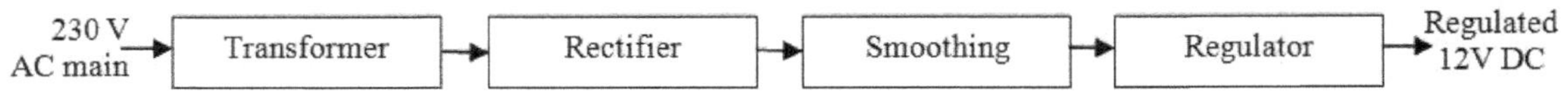

Figure 8: Block diagram of a regulated power supply system

3.3 Zero crossing detector & pulse generator circuit

This part considered as the important part of Scheme 1, where mainly the design require to put in considerations the design criterion of the zero crossing detector unit, the firing angle delay and the input control signal. The IC used is the phase control IC (TCA785) which is shown in figure 9. The synchronization signal is obtained via a high-ohmic resistance (R_5: 220k) from the line voltage (voltage V_5: 230V AC). A zero voltage detector evaluates the zero passages and transfers them to the synchronization register. This synchronization register controls a ramp generator, the capacitor C_{10} of which is charged by a constant current (determined by R_9). If the ramp voltage V_{10} exceeds the control voltage V_{11} (triggering angle α), a signal is processed to the logic. Dependent on the magnitude of the control voltage V_{11}, the triggering angle α can be shifted within a phase angle of 0 deg to 180 deg, and this can be fulfilled by using a potentiometer. For every half wave, a positive pulse of approximate 30ms duration appears at the outputs Q1 and Q2. The pulse duration can be prolonged up to 180 deg via a capacitor C_{12}. If pin 12 is connected to ground, pulses with duration between α and 180 deg will result the outputs Q1' and Q2' supply the inverse signals of Q1 and Q2. A signal of α +180 deg which can be used for controlling an external logic is available at pin 3. A signal which corresponds to the NOR link of Q1 and Q2 is available at output QZ (pin 7). The inhibit input can be used to disable outputs Q1, Q2 and Q1', Q2'. Pin 13 is used to extend the outputs Q1'and Q2'to full the pulse length (180 deg ~ α).

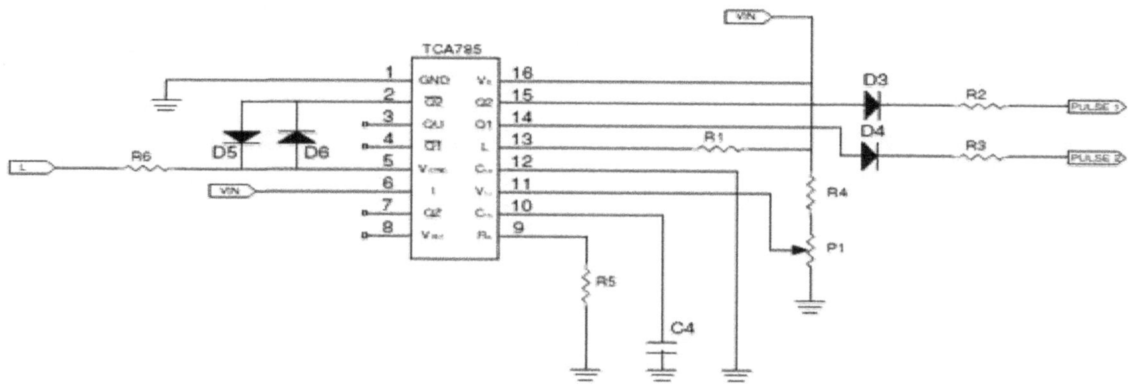

Figure 9: Pulse generator circuit

The anti-parallel diodes D5 and D6 are used to measure the two output inverters using only two pins, as at the first, the circuit will sources current through diode D6, since diode D5 is reverse biased after the current passes through D6, the D6/D5 selection signal will switch the direction of the current and force it to pass through diode D5 with D6 now is as reverse biased. This technique reduces board routing complexity, while maintains high accuracy and minimizing pin count.

3.3.1 Zero crossing detector:

These points below explain the function of the Zero Crossing Detector :

- A zero crossing detector gives a pulse when the voltage reaches 0V.
- It detects the zero voltage at pin 5, therefore whenever there is a zero voltage at pin 5, a pulse is then transferred to synchronization register.
- The synchronization register controls a ramp generator, which a ramp capacitor (C4) charges a constant current. The constant current is determined by a ramp resistor (R5).
- If the voltage of pin 10 is larger than the voltage in pin 11, the triggering angle can be shifted within a phase angle of 0° to 180°.

Figure 10 shows the zero crossing waveform signals.

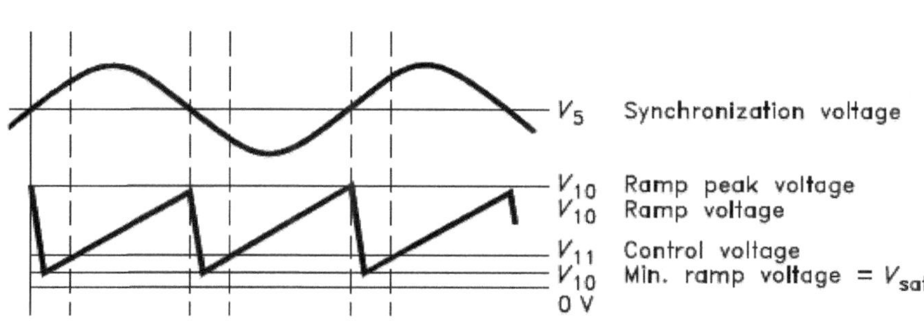

Figure 10: Zero crossing waveform signals

3.4 Pulse Transformer Circuit

Pulse transformers are used to couple a trigger-pulse generator to a thyristor in order to obtain electrical isolation between the two circuits, and also their applications in triggering considered as an excellent application. Figure 11 is illustrates the complete circuit of the pulse transformer.

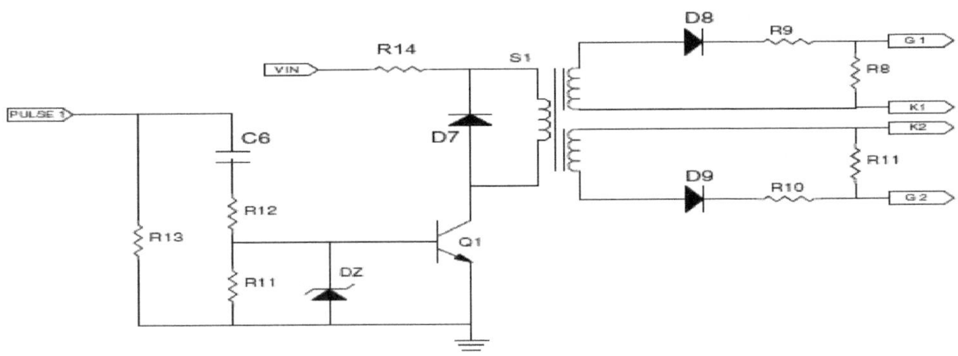

Figure 11: Pulse transformer circuit

3.5 UJT circuit for thyristor firing

The uni-junction transistor (2N2646), abbreviated UJT, is a three–terminal, single–junction device. The three terminals are: Emitter (E), Base-one (B_1) and Base-two (B_2). Between B_1 and B_2 the uni-junction has the characteristics of an ordinary resistance. This resistance is the inner-base resistance (R_{BB}) and at 25^0 C has values in the range from 4.7 k to 9.1 k. In general, R_{B1} is limited to a value below 100 ohms, and the resistor R is limited to a value between 3 k and 3 Meg. The lower limit and the upper limit of the resistor R is set by using the equation 4:

$$\frac{V_{BB}-V_v}{I_v} \prec R \prec \frac{V_{BB}-V_p}{I_p} \qquad (4)$$

These limits on the resistor R are set by the requirements that the load formed by the resistor R and V_{BB} intersect the emitter characteristic curve of Figure 12, hence the emitter peak point must be greater or smaller than I_p for the UJT to turn ON and to turn OFF respectively. In fact the value of R_{B1} is not important for the purpose of the design, provided that it is within the limits required for the UJT to oscillate. The recommended range of the supply voltage V_{BB} is 10 V to 35 V. For the design criterion of UJT, it is recommended to use a resistance of 100 ohms or greater and with large value of Wattage in series with either the power supply or in series with Base two (B_2) for a purpose of protection to the UJT from possible thermal runaway.

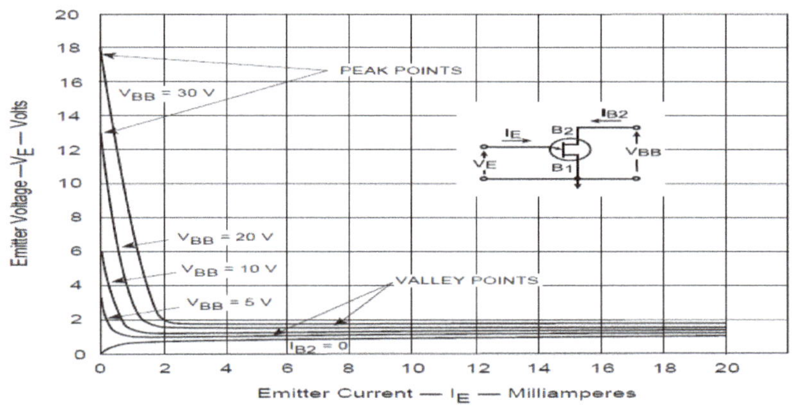

Figure 12: The emitter characteristic curve

Normally the value of the resistance R_{B2} is depending on the temperature range of the UJT, as if an extreme temperature composition is not essential, it is recommended to choose $R_{B2} \geq 100$ ohms. In addition of that, an important design consideration in this type of circuit concerns on the premature triggering of the SCR. The voltage at Base one (B1) when the UJT is turned OFF can be calculated by equation 5:

$$V_{B1}(off) = \frac{V_{BB} - R_1}{(R_{BB} + R_1 + R_2)} \qquad (5)$$

3.6 Separately- excited dc motor (Power Circuit):

This circuit is the final circuit considered as the load, it contain the bridge rectifier unit, the thyristor SCR and a 300W separately-excited dc motor. The armature winding must be connected as in figure 13.

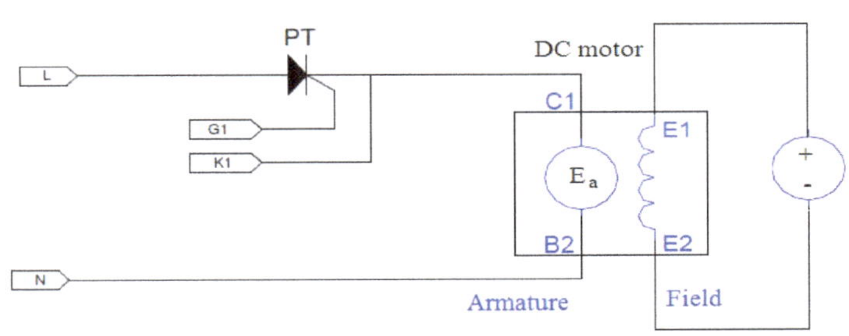

Figure 13: Power circuit of the dc motor

As C1 and B2 represent the armature winding of the motor and E1 and E2 show the connection of the field winding with the bridge rectifier unit. G1 and K1 are the outputs of the pulse transformer (secondary winding) as they represent the gate and the cathode terminals respectively. In general the field windings are used to excite the field flux, while the armature current is mainly supplied to the rotor via commutator and brush for the mechanical operation. When a separately excited motor is excited by a field current of I_f and an armature current of I_a flows in the circuit, the motor will develop a back EMF and a torque to balance the load torque at a particular speed. Normally the field current I_f is much less than the armature current of I_a. The speed control of the motor can be obtained from the firing angle α, by changing the firing angle; variable dc output voltage can be obtained. The firing angle can be varied from 0 deg to 180 deg through a potentiometer.

4.0 Experimental and test results:

Voltage Regulator output voltage using experimental calculation is 12V while output voltage during testing is 11.79V. The test result is slightly different from the experimental result mainly is because there is a voltage caused by the internal resistance of the voltage regulator device.

Pulse 1 and Pulse 2 output voltage:

- **Experimental result:**

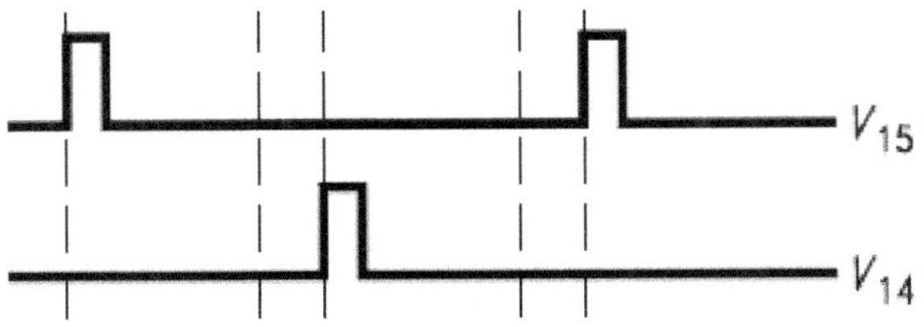

Figure 14: Expected 'pulse 1' (V15) and 'pulse 2' (V14) output waveform

- **Test result:**

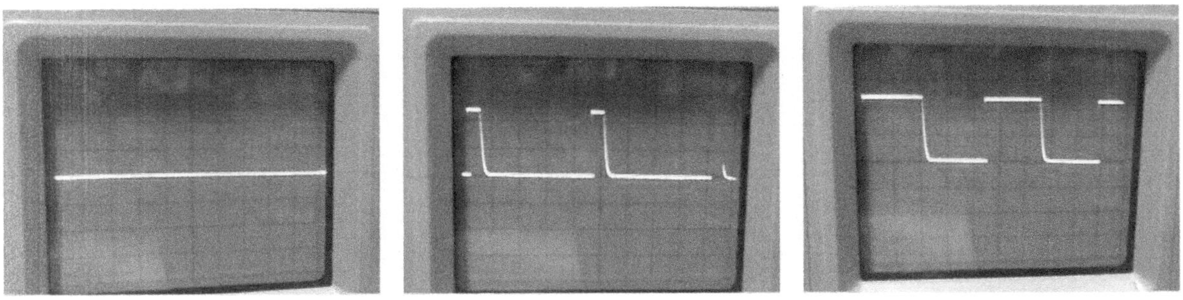

(a) Neutral line (b) Pulse 1 (c) Pulse 2

Figure 15: The actual pulse 1and pulse 2 output waveforms

For the pulse transformer, the results was obtained for the gate current for each secondary winding, and the correspond waveform. This was obtained by using the oscilloscope. For the output voltage waveform of the pulse transformer, G_1 is illustrated bellow in figure 16 (a), and the waveform of the gate after connecting the thyristor SCR, in figure 16 (b).

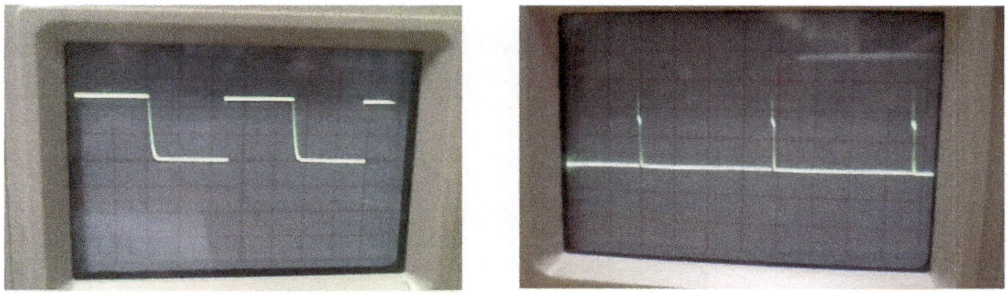

(a) Output voltage of G_1 (b)Output voltage of G_1 after connecting the SCR

Figure 16: Output voltages

5.0 The temperature controlled fan for cooling

The design of an **8051 based controlled fan** to control the speed of a fan and it has two modes, one to control the speed manually and the other is Auto in which the speed will depend on the temperature of the surrounding environment.

The system contains a sensor, ADC, Microcontroller, common cathode LED display, switches, fan and some other electronic components and the circuit was constructed as shown:

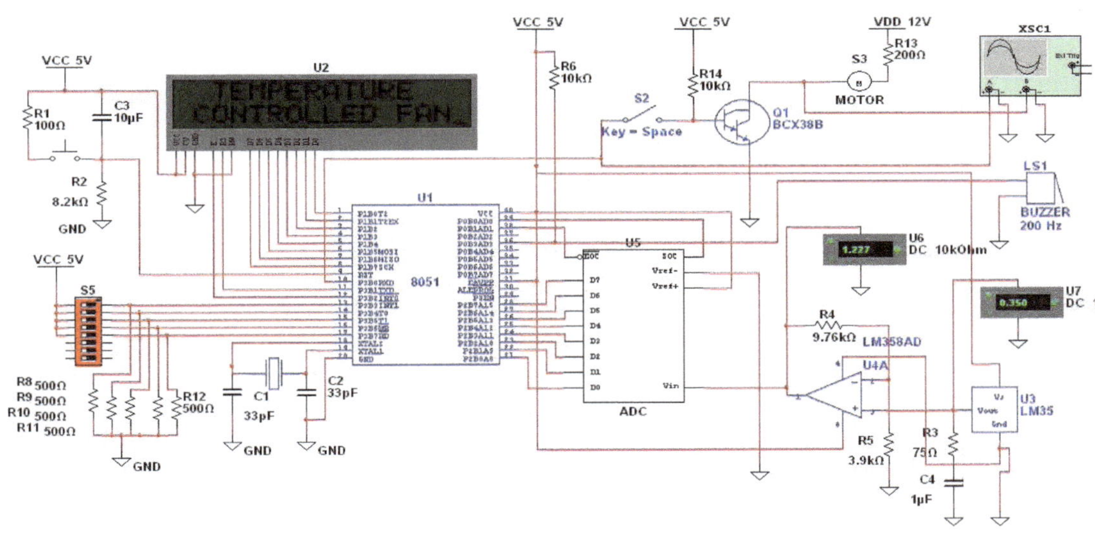

Figure 17: The temperature controlled fan schematic diagram

The sensor will detect the temperature and send it to the ADC as an analog signal. The ADC will convert the analog signal into digital signal that will be the input data of the microcontroller. This data is converted to BCD to be displayed in the LCD display as the temperature in degree Celsius. This operation will be done in both modes.

There are four switches connected to the microcontroller excluding the reset switch SW1. The switch SW2 is used to switch the system to Auto mode. The switches SW3 & SW4 are used to control the speed of the fan manually, where SW3 is to set the status of the fan to high speed while SW4 is to set it to low speed. Finally, the switch SW5 is used to switch off the fan.

In the Auto mode, the data fetched from the ADC is converted to hexadecimal data. The microcontroller then will check the temperature and change the status of the fan depending on the temperature:

Table 1: Temperature ranges and mode of operation

Mode	OFF	Low speed	High speed
Temperature range	Temperature<30°C	30°C ≤ Temperature <40°C	Temperature ≥ 40°C

6.0 Conclusion

It can be concluded that, the speed of dc motors can be controlled by using a thyristors, which is recognized as the full controlled rectifier and the design of our project meets the requirements in term of size and cost effectiveness. Moreover, some additional enhancements were made to the main circuit. Although what was done previously, some of the obstacles were faced during the designing and the construction of the circuit, such as:

- Time management.
- Technical faults (short circuit) & burning some components.
- Lake of components provided at the beginning.
- Connecting the circuit and verifying from the results.

Taking all these points into account, the overall output was a full build circuit with its own enhancement, which achieved the objectives of the project.

References

[1] Muhammad H. Rashid, 2003 ,Power Electronics: Circuits, Devices, and Applications, 3nd Edition, Prentice Hall, USA.

[2] Kothari. D P and I J Nagrath, 2004, "Electric Machines", 3rd edn, Tata McGraw-Hill, New Delhi, page no. 375 (7.14 – Speed Control of DC Motors).

[3] P.M.Balasubramaniam , 2011, "Implementation of High Power Dc-Dc Converter and Speed Control of Dc Motor Using DSP", Kalaignar Karunanidhi Institute of Technology, Computer Engineering and Intelligent Systems, Vol 2, No4.

[4] Dr.Bimbhra.P.S. and Khanna Publishers, 1998, "Power Electronics", New Delhi, 2nd Edition.

[5] Moleykutty George, 2008, Speed Control of Separately Excited DC Motor, *American Journal of Applied Sciences,* Volume 5, Issue 3, Pages 227-233.

[6] Liu, Kwang-Hwa and Lee, F. C, 1990, "Zero Voltage Switching technique in DC/DC Converters", IEEE Transactions on Power Electronics, page no. 293 – 304.

[7] G. Rajeshkanna, 2013, "Modern Speed Control of Separately Excited DC Motor by Boost Converter Fed Field Control Method", International Conference on Computer Communication and Informatics, Pages 1 – 7.

[8] Jianguo Zhou, et al., 2001, "Global speed control of separately excited DC motor" , IEEE CONFERENCE PUBLICATIONS, Volume: 3, Pages 1425 – 1430.

[9] R.Krishnan, 2005, "Electric Motor drives modeling, Analysis and control", Prentice-Hall of India Private Limited, New Delhi.

[10] Data sheet catalog Homepage, Function and Industrial ICs - Phase Control IC (TCA785) [online]. Available: http://www.datasheetcatalog.com/datasheets_pdf/T/C/A/7/TCA785.shtml.

[11] T.C. Burg, D. M. Dason, J. Hu, P. Vedagarbba, 1994, "Velocity tracking for a separately excited dc motor without velocity measurement", Proceeding of the American Control Conference, pages 1051-1055.

[12] Hsu, Yuan-Yih, Chan, Wah-Chun, 1984, "Optimal variable-structure controller for DC motor speed control", IET JOURNALS & MAGAZINES, Volume: 131 , Issue 6 , Pages 233 – 237.

14

Design and Analysis of Column Mounted JIB Crane

K Suresh Bollimpelli[1], V Ravi Kumar[2]

[1]*ASR College of Engineering & Technology, bollimpelli.ksuresh@gmail.com*

[2]*ASR College of Engineering & Technology, vadlamudi717@live.com*

Author Correspondence: K Suresh Bollimpelli, Flat No: 104 Spurthi Towers Tanuku.PIN 534211

Abstract

In this work, a static, modal and harmonic analysis of a column mounted jib crane using ANASYS software is presented. A column mounted jib crane of 1.5 Ton capacity is modelled using CATIA which is imported into ANASYS where calculations are performed. The detailed drawing of various parts of the crane is obtained from TATA Advanced systems Ltd (TASL) Adibhatta village, Hyderabad. The deflection values, Von Misses stress etc are obtained using the static analysis. The hand calculations of the column mounted jib crane have been done using simple strength of material expressions. The deflection is obtained as 3.709mm, when the load applied is 1.5 tons. The maximum stress obtained is 147.8Mpa which is less than the allowable stress. The static stress was found to be within the limits of safety. The model analysis shows the natural frequencies of the crane to be in the lower range 0-10Hz. The fundamental frequency is found out to be 0.323589 Hz. All the other higher frequencies are also found to be very low making the jib crane less stiff and highly stable for any transient loading. The harmonic analysis is performed with a view to predict the performance of the crane if a cycle time dependent load is allowed to act at the trolley. For this hypothetical situation, the von-mises stress and displacement along the z-directions were obtained using ANSYS. The maximum von-Mises stress of 60Mpa occurs at fundamental frequency of 1 Hz. The maximum z-direction displacement of 5mm was observed. These values indicate that the column mounted jib crane is safe to operate under the load of 1.5 Tons under static and cyclic time dependent loads also

Keywords: JIB CRANE, ANSYS, CATIA.

1. Introduction

A jib crane is in effect a monorail that is cantilevered from its supporting members and pivoted at one end. The horizontal beam provides the track for the hoist trolley. Jib crane have three degrees of freedom. They are vertical, radial, and rotary. However they cannot reach into corners. They are usually used where activity is localized. Lifting capacity of such cranes may vary from 0.5 ton to 200 ton and outreach from a few meters to 50 meters. Such cranes find various applications in port area, construction site and other outdoor works. For handling general cargo, lifting capacities usually 1.5 ton to 5 ton with maximum out reach of 30 meter. Jib crane provided with grabbing facilities have usually a capacity ranging from 3 ton operating 50 to 100 cycles per hour. Lifting heights may be 30 meter or more. Jib cranes used in ship yards for lifting heavy machinery equipment, weighing 100 to 300 tons, are usually mounted on pontoons [1]. Frequently, these cranes are provided with two main hoisting winches which can be employed singly or together to lift a load. For handling light loads may hand auxiliary arrangement localized, such as in machine shops. Column mounted jib cranes are commonly used in packaging industry. The size of the crane can be visualized from the height of the operator. These cranes are used for hoisting up to 1 ton loads.

2. Material type

The material is considered is structural steel ASTM A36 steel [2] in this work, static, model, and harmonic analysis of an EOT using ANSYS has been performed. The model was prepared using CATIA software, and imported to ANSYS. The model is then meshed using 3-D solid elements (Solid45).

3. Manual Design of the parts of a JIB Crane

A column mounted jib crane consists of following distinct part, viz., (i) jib (ii) Post (iii) Base plate (iv)Tie rod. Figure (1) shows the various parts of the jib crane .the jib crane consists of a base which is fixed to the ground at the bottom. The bottom side of post is connected to the ground with the help of a fixed support. The motion of the trolley is derived from an electric motor mounted in the trolley itself. The wheels run on the flanges of the 'I' sectioned jib along the length of the jib.

The trolley consists of hoisting machinery which raises or lowers the load by connecting it with a hook. The load hook has three separate motions, these beings the hoisting, longitudinal traverse of the trolley, and swiveling of the crane through 180 degrees. Each motion is controlled independently of the other motions by separate controllers situated in a control cage or in a suitable position for controlling from the floor by pendant chains.

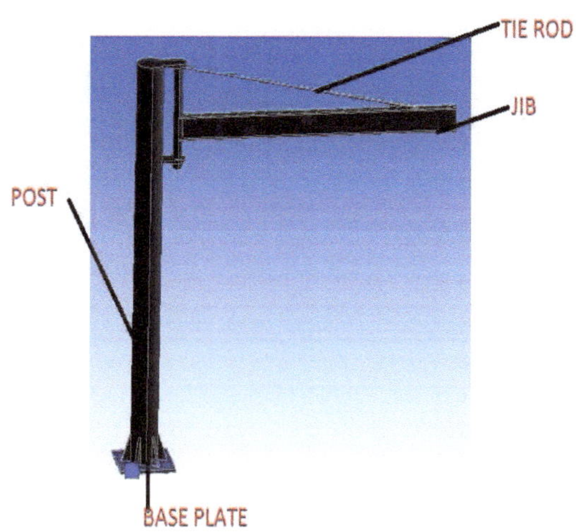

Figure 1: Various parts of a Column Mounted JIB Crane

3.1 Design of Tie rod:

The tie is inclined at an angle of 30^0 to the jib and is attached at a point 2.5 meters from the center line of the crane post, thus allowing a clear radius of 2.5 meters from the buckle, for facility in erection as well as to take up any adjustments due to faulty workmanship. Also it is assumed that no fixing moment is exerted on the jib by the crane post-in any case such moment would be extremely small in this design. The maximum load in the tie will occur when the hook block is at the extreme of our position i.e At 2.5 meters radios. Drawing the triangle of forces we get tension in the tie as=30,000 N. and compression in the jib=25980 N approx

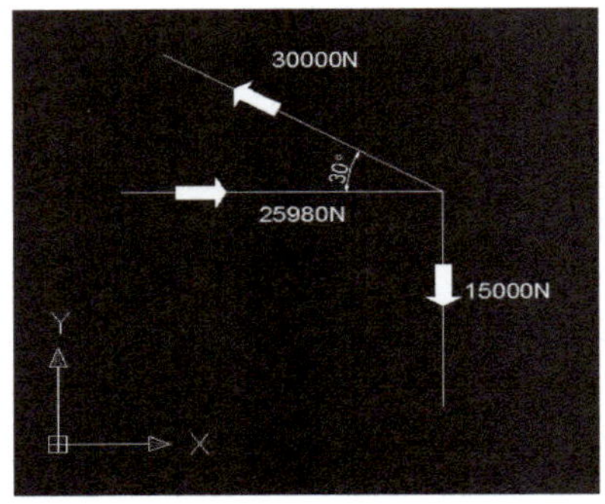

Figure 2: Resolution of Forces on the Tie-Rod (using AUTOCAD)

The weights of jib, tie rod and the trolley have been neglected for the time being actual loads will be checked later on when on some dimensions have been assigned to these parts. A round tie bar of M.S having the upper end forged into the shape of an eye for pin joint is proposed for attachment to crane post. A maximum permissible tensile stress of about 60 N/mm^2 would be allowed in all structural members. In slow motion hand cranes an important allowance of 15 to 20% must be allowed

In other words, all the actual working loads and stresses in various members of the structure should be increased by 30 to 40% to get their equivalent static values. A factor of safety of 4 should be allowed then on these static values on basic of ultimate stress to get the maximum permissible working stress. Suppose, actual working load in certain members is 1000 N. Then its static value after allowing 40% impact factor would be 1200 N. Tenacity of M.S=350 N/mm^2 .approx(static), or 350/(4*1.4)=62.5 N/mm^2 (dynamic), if the calculations are not complicated due to the introduction of the "impact factor"

Cross sectional area of the bar required=30000/62.5

$A=480\ mm^2$ approx.

$A=480\ mm^2$

$$(\Pi/4)*D^2 = 480\ mm^2 \quad (1)$$

Therefore D=24.72 mm, D=25 mm,

This diameter is at the bottom of the thread.

Core diameter =25 mm,

Full diameter =25/0.85 mm, say 29.41mm,

Therefore the diameter of the tie rod required =30 mm.

3.2 Design of JIB

The maximum bending moment on the jib occurs when the load is at the free end of the jib which is 2.5 m from the fixed end. The shearing force will be 10000N only and may not be taken in to consideration.

Max Bending Moment

$$= WL = 150000\times2500 = 375,00,000\text{N-mm} \quad (2)$$

Allowing a maximum permissible bending stress of 250N/mm^2.

$$\text{Section modules of jib required} = \frac{37500000}{165} \quad (3)$$

$$=227272.72\text{mm}^3$$

A rolled steel joist will be most suitable for this design. The load trolley will run on the inner tapered surface of the lower flange.

In the design of electric and hand operated cranes and crane gantries and runways etc. it is most important to check the deflection of the beam and girders so that they do not exceed certain prescribed limit. If the deflection of a girder (on which the trolley runs) is more than prescribed limit the trolley will be obstructed in their passage or probably inclination of the track to the horizontal will be increased, there by more power(or force) will be required to drag the load trolley along the girder length. This will also create a unpleasant jerky or surging B.S.466, clause 18, restricts the deflection of a crane girder at the point of its Max. B.M. to the following limits.

For bending stress intensity of $124N/mm^2$.

The ratio of span/girder depth i.e, $\frac{L}{d}$ (4)

Should be 7.67 mm to get a maximum deflection of $\frac{SPAN}{325}$. [3] (5)

A reference to ISI hand book for structural engineers; "structural steel section " shown that depth of nearest std. R.S beam is either 300mm or 400mm. either a 300mm × 165mm section or 400mm×165mm. section, the moment of inertia I_{XX} being $8603.6cm^4$ and $13630.3cm^4$ respectively. The section modulus Z_{xx} in each case will be 8603.6/17.5 and 13630.3/20 i.e. 752 and 965 respectively. But we do not require a section modulus of more than 17.5 cm^4 if such bulky sizes of 400×165mm; or 350×165 mm, girders are used, They will not only be wasteful of material but will also increases the manufacturing costs unnecessarily and use less weight to the component of the crane. Try a 225mm×100mm. R.S. Beam and limit the compressive bending stress intensity to $63N/mm^2$ approx. so that the ratio of girder length/width of compressive flange should not exceed 45:1 in the present case the ratio will be 250/10 =25:1 which is O.K.

And, maximum permissible compressive stress

$F_c = F_t\,[1\text{-}0.1\times (I/b)]$ (6)

$=165[1-(0.1\times0.25)]$

$=124\ N/mm^2$.

The section modulus required at $124N/mm^2$

stress = 375, 00,000/124

$=302419.35mm^3$ approx

The actual section modulus Z of a 225mm×100mm R.S girder having a weight of 23.5kgf/meter length 2502/10=250000 approx checking the deflection, it will be observed that the ratio of I/d, for a stress intensity of $124N/mm^2$ should be 13.0 to given a deflection of span/1500, actual ratio of I/d, if a 350mm×100mm R.S beam is used, is 300/22.5=14.7. This ratio will serve our purpose; therefore adopt 350mm. 100mm, 23.5kg/meter length ISMB 350 R.S beam for the jib .when the trolley is at the outer end of the jib, the jib will be in the compression under a stress of 25980N, and under a shearing force of 15000N. The cross sectional area of the beam being 6671 mm^2 (large enough). The compressive or shear strength or principle stress need not be investigated.

The deflection of the jib a free end is given as $\Delta = \frac{WL^3}{3EI}$ [3] (7)

Where W=load acting on the jib =15000N

L=length of the jib =2500 mm.
E=young's modulus of jib material $=2\times10^5$ N/mm^2.
I=moment of inertia of jib section =1363030000 mm^4.
Know substitute this in the formula
Then Δ =2.86 (Apporx).

3.3 Design of the post:

The length of the crane post between the upper brackets and fixed support act as a beam, and the tension acts as an inclined concentrated load. The beam is subjected to the loading as shown in fig.3.3.1 taking moments about bearing bracket A,

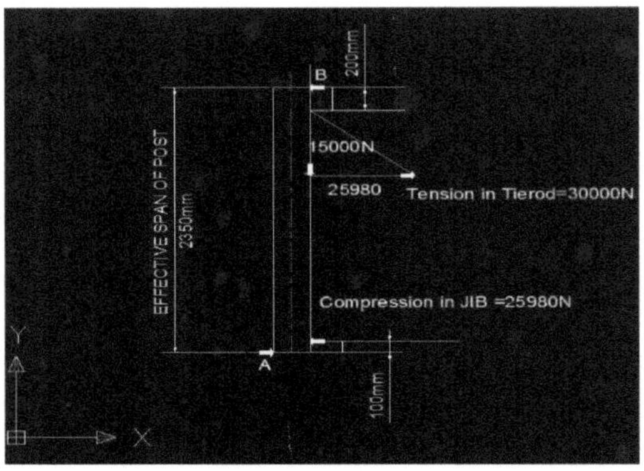

Figure 3: various forces acting on the post (using AUTO CAD)

(25980×10)+ (B×235)=25980×215.0 (8)
B=236583.82N approx
Similarly taking moments about bearing bracket B,
(25980×20)+(A×235)=25980×225 or
A=(25980×225)-(25980×20)/235
Reaction force at A=22663.40 N approx.
Max B.M negative at L.H side
=23658×10=23658N-cm.
Max B.M positive at R.H side
=22663×20=45326N-cm.
=532600N-mm.
Allowing a max permissible stress of 94N/sq-mm,
Section modulus Z of crane post required =45326/940
=48.22cm^3 approx
=48220 mm^3

For a CIRCULAR cross section

$$Z=\frac{\pi}{32}*60^3 \quad = \quad 21205.75\text{mm}^3$$

Moment of inertia of post$=\frac{\pi}{64}*60^4$ =636172.51mm^4

Designing of the post for deflection

Max deflection= $\frac{2\sqrt{2}P'L3}{27^2 \times \sqrt{3} \times EI}$ = 4.56MM (9)

The nearest standard size of circular cross section of diameter 60 mm is used. In addition to the bending stress, a direct compressive stress or thrust is induced in the crane post whose value is 1.5tons (1500 kg).
Cross sectional area of 60 mm, circular rod=2827.4 mm^2

Actual stress intensity due to direct thrust=15000/2827.4 (10)

=5.3051N/mm^2 (approx)

Actual bending stress intensity =4532680/21205.75 (11)

=213.74/mm^2 (approx)

Total stress in the crane post

=213.74+5.305=219.05 /mm^2 (approx) will do.

The diameter of the crane post within the bearing may be reduced to 65 mm. The maximum load on each bearing is 236583N approx. allowing a bearing length of 1 $\frac{1}{2}$ times the journal diameters, the length of each bearing adopted 65×1.5=98, say to 100mm.

Bearing pressure= $\frac{23658}{65 \times 100}$ =3.6396N/mm^2 of projected area. (12)

Though the barring pressure appears to be high, yet the motion of slewing will be slow and very intermittent, and for this reason it may be allowed provided suitable oil holes or grease cup or nipples are allowed for in the bearing brackets..

3.4 Design of upper bearing bracket:

The reaction of the bearing 23658N will acts as shown in opposite. This force produces a direct horizontal tensile stress of 23658/4=5914.59N. Approx per bolt using an 18mm diameter bolt, the maximum tensile stress intensity = $\frac{5914}{(18)^2 \times 0.7854}$ =23.24N/mm^2 approx (O.K) (13)

The rest of dimensions of C.I bearing brackets as show opposite can be developed the drawing sheet.

3.5 Design of lower bearing bracket

The horizontal reaction in this case acts in the opposite direction: i.e towards the wall column there is therefore, the tensile stress induced in the bolts due to the vertical thrust of 20000 on the bracket each will be under a direct shear of 20000/4=5000N.

Shear stress intensity = $\frac{3750}{0.7854 \times 18^2}$ (14)

=14.73N/mm^2. Approx (O.K)

As the size of G.M bearing bush is exactly the same as that top for the top bearing bracket the same casting can be used for the lower bracket.

4 CATIA Model of JIB Crane

Column Mounted JIB Crane model produced using CATIA V5 R17 [4]

Figure 4: Final assembly of column mounted jib crane

5 Results and Discussion

5.1 Static Analysis

Importing of CATIA V5R17 file into ANSYS workbench R14.5 [5]

Model (A4)

Geometry:

Table 1: Model (A4) > Geometry

`	*Geometry*
State	Fully Defined
Definition	
Source	C:\Users\ome\Desktop\part123.igs
Type	Iges
Length Unit	Meters
Element Control	Program Controlled
Display Style	Body Color
Bounding Box	
Length X	0.6 m
Length Y	3.6663 m
Length Z	3.705 m
Properties	
Volume	0.29885 m^3
Mass	2331.1 kg
Scale Factor Value	1.
Statistics	
Bodies	1

Active Bodies	1
Nodes	64968
Elements	35325
Mesh Metric	None
Basic Geometry Options	
Solid Bodies	Yes
Surface Bodies	Yes
Line Bodies	No
Parameters	Yes
Parameter Key	DS
Attributes	No
Named Selections	No
Material Properties	No
Advanced Geometry Options	
Use Associativity	Yes
Coordinate Systems	No
Reader Mode Saves Updated File	No
Use Instances	Yes
Smart CAD Update	No
Attach File Via Temp File	Yes
Temporary Directory	C:\Users\ome\AppData\Local\Temp
Analysis Type	3-D
Mixed Import Resolution	None
Decompose Disjoint Geometry	Yes
Enclosure and Symmetry Processing	Yes

Mesh:

Table 2: Model (A4) > Mesh

Object Name	*Mesh*
State	Solved
Defaults	
Physics Preference	Mechanical
Relevance	0
Sizing	
Use Advanced Size Function	Off
Relevance Center	Medium
Element Size	2.e-002 m
Initial Size Seed	Active Assembly
Smoothing	Medium

Transition	Fast
Span Angle Center	Coarse
Minimum Edge Length	1.4013e-003 m
Inflation	
Use Automatic Inflation	None
Inflation Option	Smooth Transition
Transition Ratio	0.272
Maximum Layers	5
Growth Rate	1.2
Inflation Algorithm	Pre
View Advanced Options	No
Patch Conforming Options	
Triangle Surface Mesher	Program Controlled
Advanced	
Shape Checking	Standard Mechanical
Element Midside Nodes	Program Controlled
Straight Sided Elements	No
Number of Retries	Default (4)
Extra Retries For Assembly	Yes
Rigid Body Behavior	Dimensionally Reduced
Mesh Morphing	Disabled
Defeaturing	
Pinch Tolerance	Please Define
Generate Pinch on Refresh	No
Automatic Mesh Based Defeaturing	On
Defeaturing Tolerance	Default
Statistics	
Nodes	64968
Elements	35325
Mesh Metric	None

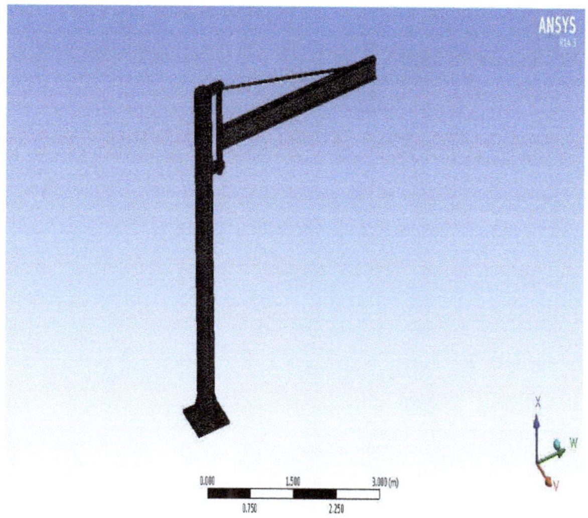

Figure 5 Meshed part of a column mounted jib crane

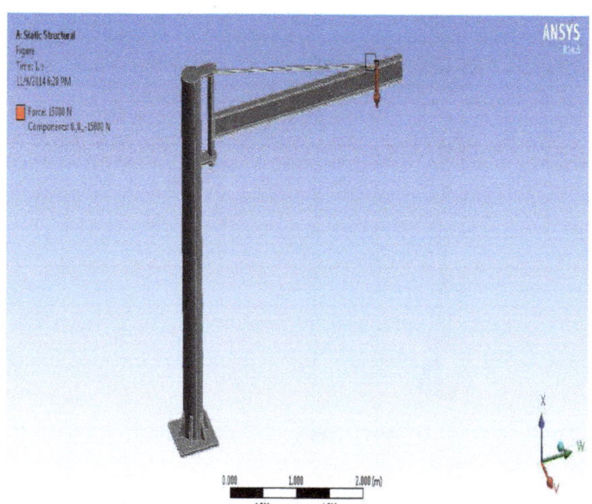

Figure 6: Load applied on column mounted jib crane

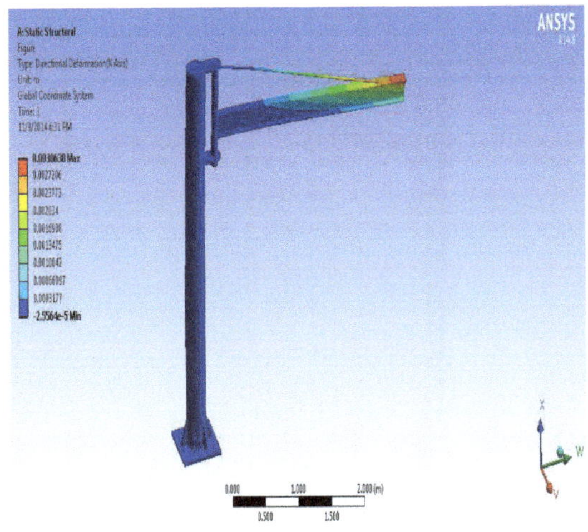

Figure 7: Displacement along the x-direction

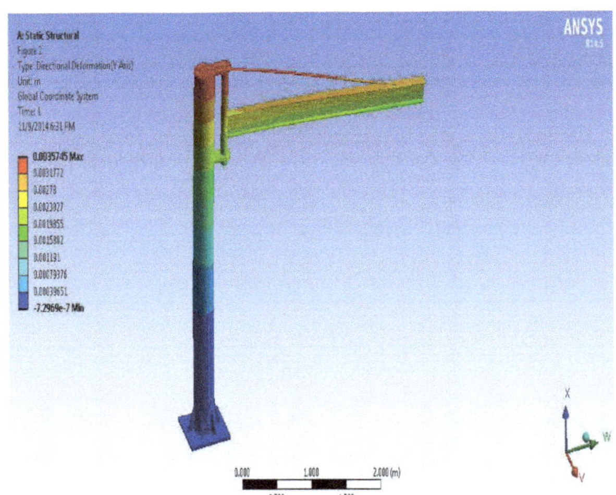

Figure 8: Displacement along the y-direction

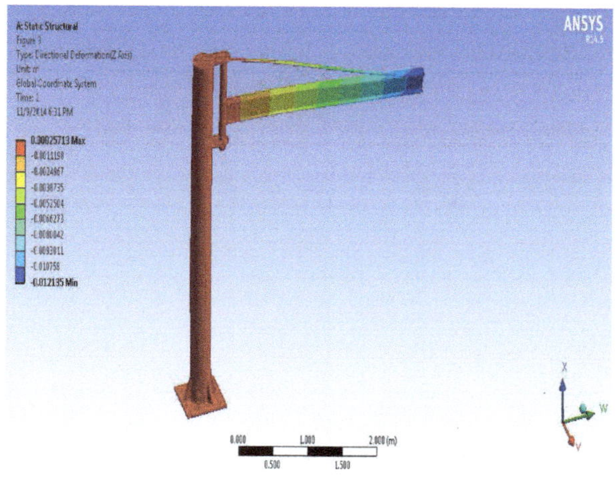

Figure 9: Displacement along the z-direction

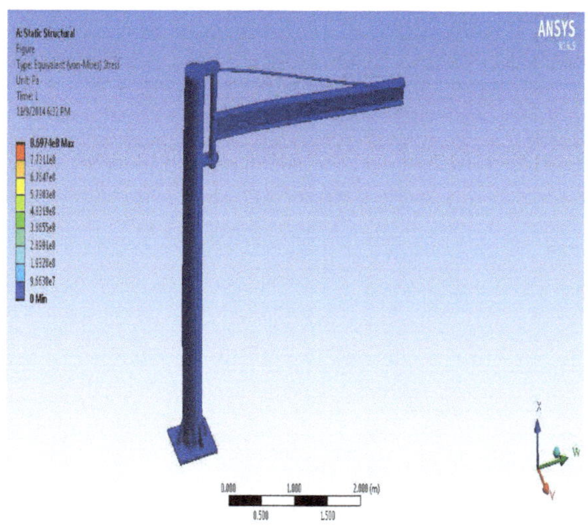

Figure 10: Von mises stress

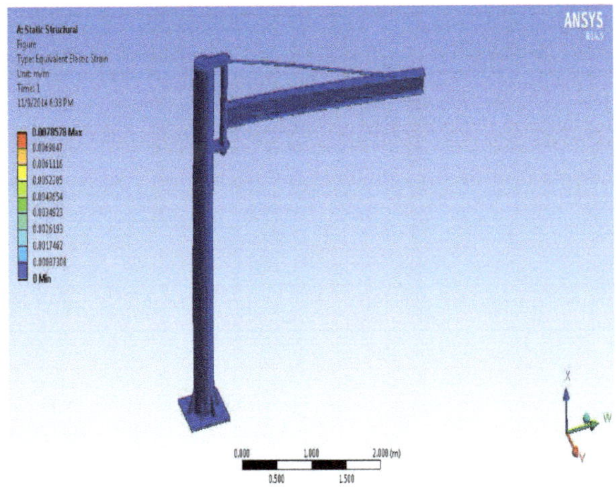

Figure 11: Von mises strain

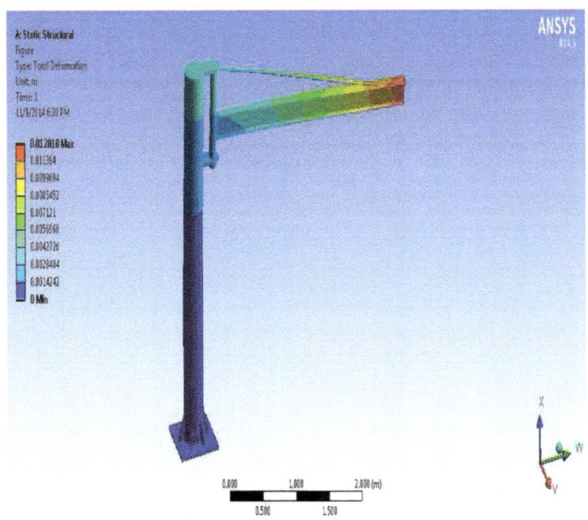

Figure 12: Total Deformation

5.1.1 Material Data

ASME A 36 Steel [2]

Table 3: ASME A 36 Steel > Constants

Density	7800 kg m^-3

Table 4: ASME A 36 Steel > Isotropic Elasticity

Temperature C	Young's Modulus Pa	Poisson's Ratio	Bulk Modulus Pa	Shear Modulus Pa
	2.e+011	0.32	1.8519e+011	7.5758e+010

5.1.2 Results:

Figure 5 to 12 shows the displacement along the x, y, and z-directions, von mises stresses and von mises strains when the load of 1.5 tons is applied at the free end of the jib. The maximum deflection can be seen to be 3.507 mm over a span of 2.5m. The displacement in the x direction is 2.72 mm, and the displacement in the y direction is 3.57mm and the displacement in the z direction is 0.25mm. The maximum von mises stress is 156.8N/mm^2 and the von mises strain is 0.00785. This is found to be acceptable for the operation of the jib crane.

5.2 Model Analysis

Importing of CATIA file in to ANSYS Workbench

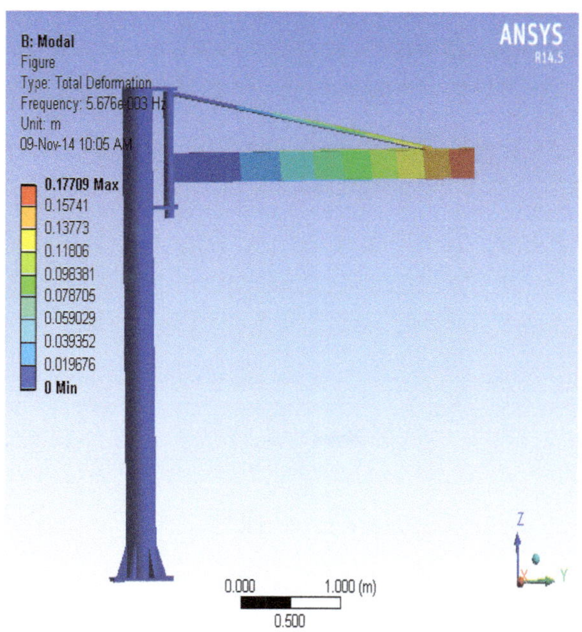

Figure 13: Model 1

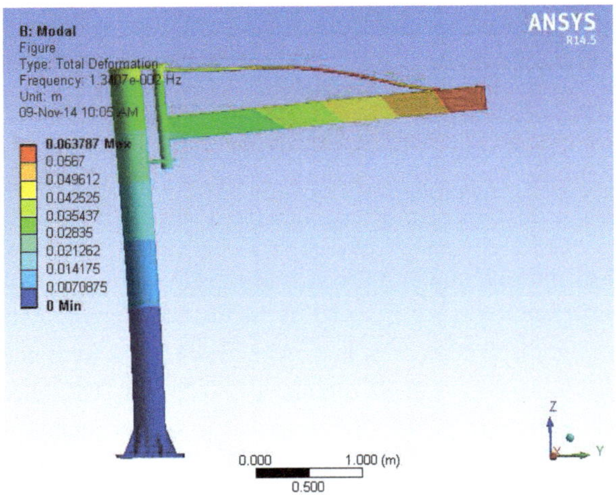

Figure 14: Model 2

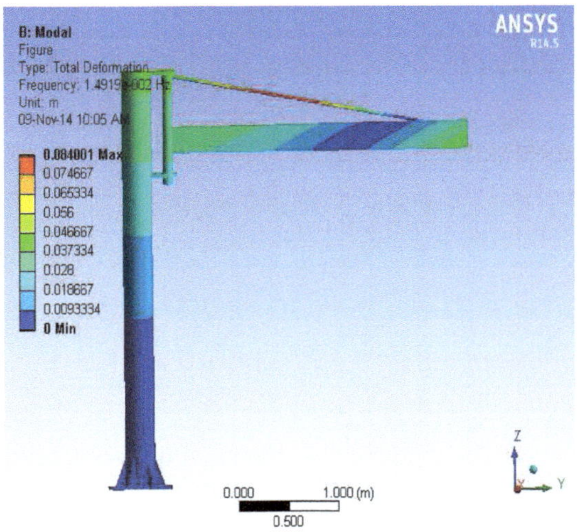

Figure 15: Model 3

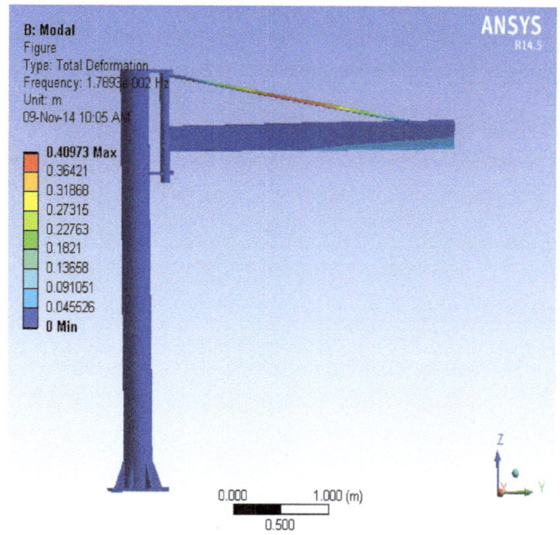

Figure 16: Model 4

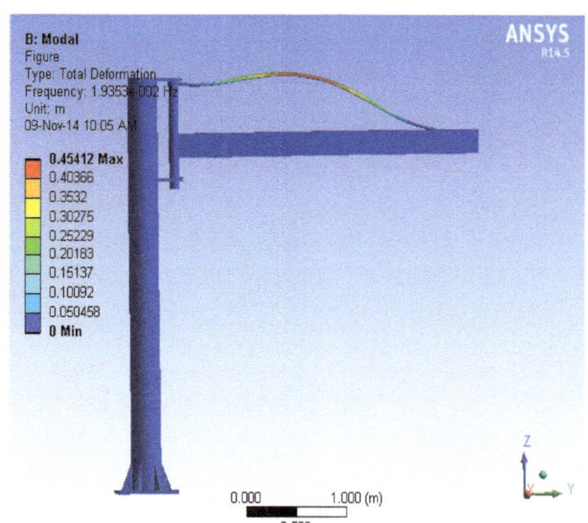

Figure 17: Model 5

5.2.1 Results:

Figure 13 to 17 shows the first five model shapes of the jib crane. For model 1the frequency is 0.2825 Hz and displacement is 0.177mm. For model 2 the frequency is 0.18Hz and the displacement is 0.63 mm. For model 3 the frequency is 0.20 Hz and displacement is 0.84 mm. For model 4 the frequency is 0.24Hz and the displacement 0.41 mm. For model 5 the frequency is 0.26Hz and the displacement is 0.41mm.These low frequencies indicate that the crane is not very stiff and hence it is stable in operation.

5.3 Harmonic Analysis

Model (C4) > Harmonic Response (C5) > Analysis Settings

Object Name	*Analysis Settings*
State	Fully Defined
Options	
Range Minimum	0. Hz
Range Maximum	10. Hz
Solution Intervals	5
Solution Method	Mode Superposition
Cluster Results	No
Modal Frequency Range	Program Controlled
Store Results At All Frequencies	Yes
Output Controls	
Stress	Yes
Strain	Yes
Nodal Forces	No
Calculate Reactions	Yes
General Miscellaneous	No
Damping Controls	
Constant Damping Ratio	0.
Stiffness Coefficient Define By	Direct Input
Stiffness Coefficient	0.
Mass Coefficient	0.
Analysis Data Management	
Solver Files Directory	C:\Users\Owner\AppData\Local\Temp\WB_OWNER-PC_3272_2\unsaved_project_files\dp0\SYS-2\MECH\
Future Analysis	None

6. Solution:

Type	Total Deformation	Directional Deformation			Equivalent Elastic Strain
Orientation		X Axis	Y Axis	Z Axis	
Coordinate System		Global Coordinate System			
Results					
Minimum	0. m	-2.6286e-003 m	-8.4173e-003 m	-4.7465e-002 m	0. m/m
Maximum	4.7628e-002 m	2.2296e-003 m	8.4786e-003 m	6.6611e-003 m	0.12174 m/m
Information					
Reported Frequency	10. Hz				

Above process is applied for the various frequencies for examples 1Hz, 2Hz, 3Hz, up to 10Hz .we got the values for various x-directional displacement, y-directional displacement, z-directional displacement, von-mises stress and von-mises strain. These values are plotted in graph.

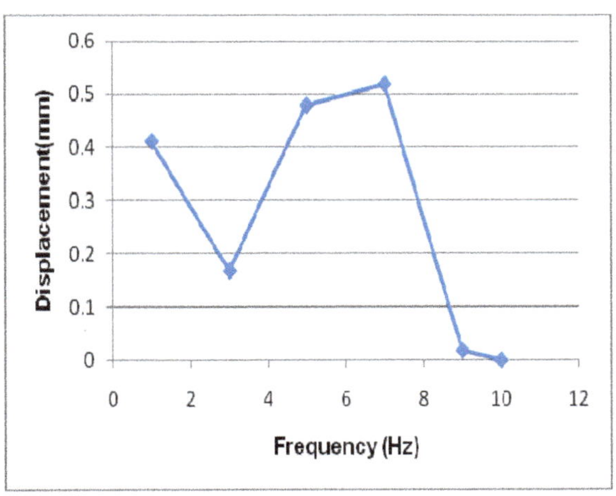

Figure 18: Displacement along the X-direction

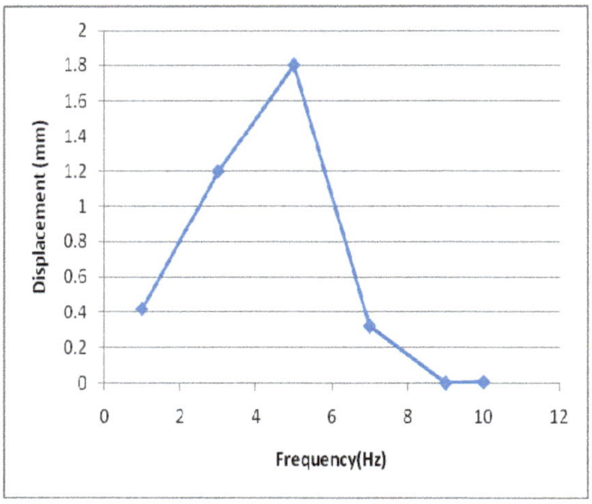

Figure 19: Displacement along the Y-direction

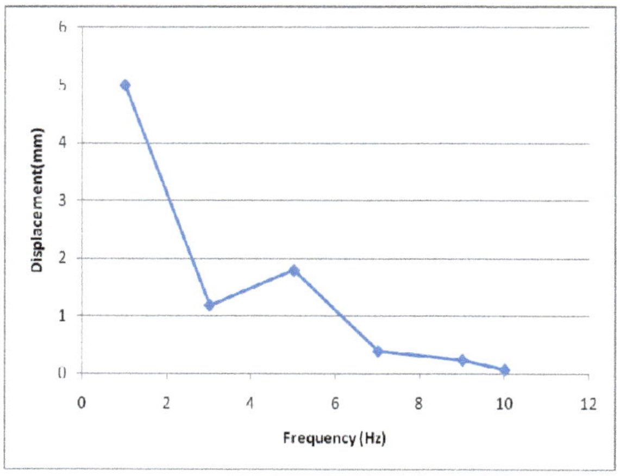

Figure 20: Displacement along Z-direction

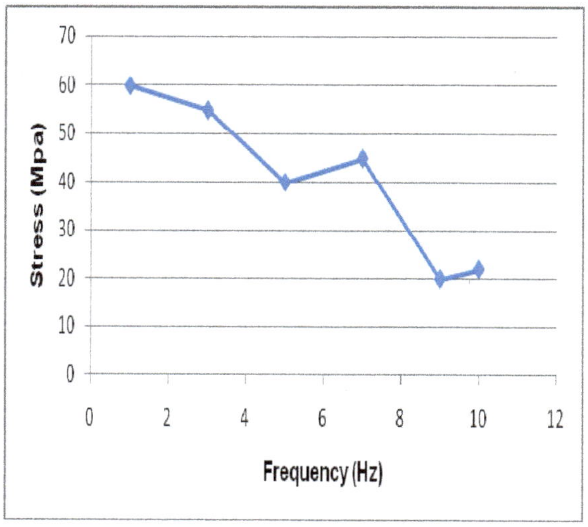

Figure 21: Von Mises stress

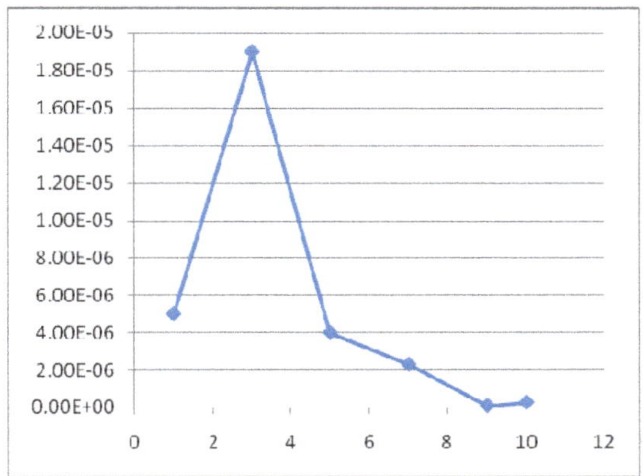

Figure 22: Von Mises strain

7. Result:

Figures 6.1 to 6.5 shows the graphs between X displacement Vs frequency, Y displacement Vs frequency, Z displacement Vs frequency, Von Mises stress Vs frequency, and Von Mises strain Vs frequency. The harmonic analysis is performed with a view to performance of the crane if a cycle time dependent load is allowed to act at the trolley. For this hypothetical situation, the Von-Mises stress and displacement along the z-directions were obtained using Ansys. The maximum Von Mises stress of 60Mpa occurs at fundamental frequency of 1 Hz. The maximum X-direction displacement of 0.52 at 7Hz and the maximum Y-directional displacement of 1.8mm were observed at a frequency of 5Hz and the maximum Z-directional displacement of 5mm was observed at a frequency of 1Hz and the maximum strain is found to be 1.9e-5 at a frequency of 3Hz. These values indicate that the column mounted jib crane is safe to operate under the load of 1.5 Tons under static and cyclic time dependent loads also.

8. Conclusion

The static, model and harmonic Analyses of a Column mounted jib crane have been performed as per the required load conditions. The model is obtained from TATA Advanced systems Ltd (TASL), Adibhatta village, Hyderabad.

The static analysis performed on jib crane yielded a maximum von-Mises stress of 156.8N/mm^2 which is the Yield stress limit of the material chosen (250 MPa). The results agreed very closely with hand calculations performed by assuming a simplified model.

The model analysis is provided information regarding the natural frequencies of the jib crane. The fundamental frequency is found out to be 0.2825 Hz and the 5th mode has a frequency of 0.26Hz. These low frequencies indicate that the crane is not very stiff and hence it is stable in operation.

The harmonic analysis is performed with a view to predict the performance of the crane if a cyclic time dependent load is allowed to act at the trolley. For this hypothetical situation, the Von-Misses stresses and displacement along the z-directions were obtained using Ansys. The maximum Von-Misses stress of 60 MPa occurs at fundamental frequency of 1Hz. The maximum displacement in the direction of the application of load is 5mm. These values indicate that the column mounted jib crane is safe to operate under the load of 1.5 Tons under static and cyclic time dependent loads also. It may be concluded that the column mounted jib crane is safe for operation up to 1.5 tons.

References

[1]. Rudenko, N., *"Material handling Equipment,"* Moscow:peace,1969

[2] *ASME Section - II* (Ferrous Materials)

[3] *Marks' Standard Handbook for Mechanical Engineers*, Eleventh Edition; by Eugene A. Avallone, Theodore Baumeister III, Ali Sadegh

[4] " *CATIA V5R17 for Designers*" by Sham Tickoo

[5] *Finite Element Simulations with ANSYS Workbench 14*, SDC publications

A Brief Author Biography

K Suresh Bollimpelli is pursuing M.Tech (CAD CAM) at ASR college of engineering and technology, Tanuku. His field of research interest is designing Gigs and Fixtures . .

REVIEW ON SOLAR FOOD CUTTER WITH SHAFT MOUNTED SPEED REDUCER FOR AGRICULTURAL FIELD APPLICATION

Nikhil S Raut[1], Ashwin Mahindrakar[2]

[1]*PG Student, Nikhilr617@gmail.com*
[2]*Assistant Professor, ashwin.mahindrakar@sinhgad.edu*
Author Correspondence: Snehankit colony, karve nagar, Pune.

Abstract

Indian agriculture has a lot of dependency on farm labour for many kinds of operation. The main products of the farm are needed to be harvested. Now-a-days the cutting of fruits; flowers and vegetables are done by snips. Hence the main purpose now-a-days is to reduce the labour hours. Thus taking into account the fact that the development of a portable, light weight multi-purpose agricultural cutter using solar powered 12volts D.C motor is required. For this purpose the development of the cutter and then the shaft mounted speed reducer gear box is to be done. After development of cutter, the stress developed at the cutting edge and force acting on entire linkage will be measured. Then these stresses will be evaluated in ansys by applying proper boundary condition. Again different material will be tested and examined for manufacturing.

***Keywords**:* Agricultural cutter, shaft mounted speed reducer, Solar panel.

1. Introduction

Agriculture is an important part of the Indian economy and culture, and it can play an important role in distributed generation of energy. This project concept identifies the opportunities for solar energy use in agriculture. Farmers have the tradition of being stewards of the land, and their investment in renewable energy supports their role of protecting the land, air, and water. Solar energy, like other renewable, offers an opportunity to stabilize energy costs, decrease pollution and greenhouse gases (GHGs), and delay the need for electric grid infrastructure improvements. Solar energy systems have low maintenance costs, and the fuel is free once the higher initial cost of the system is recovered through subsidies and energy savings (from reduced or avoided energy costs).

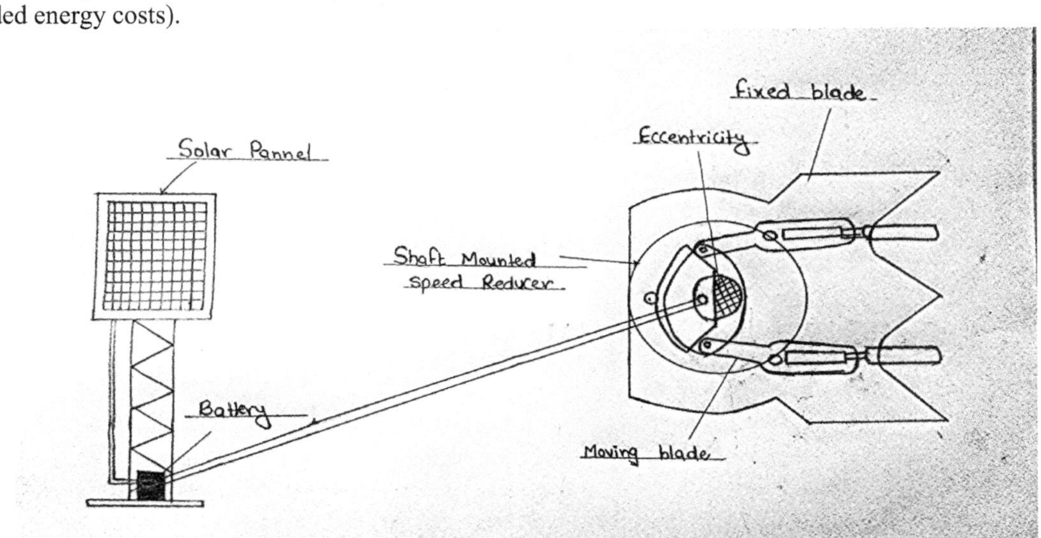

Figure 1: Solar Agricultural Cutter

1.1 Problem Statement

Indian agriculture has lot of dependency on farm labour for variety of operations to be performed in agriculture farm right from land preparation for cultivation to harvesting of the agriculture produce. Fruits, vegetables, flowers etc, that are grown in the farm need to be harvested. Meaning that they have to cut from the plant using suitable methods so that they can be packed and then transported to the market.

The cutting of vegetables like Brinjal (eggplant), lady finger, Capsicum, pumpkin, bottle gourds, bitter gourds, ridge gourds, cucumber, pumpkin, green peas, French beans, tomatoes, etc is normally done by use of plucking by fingers or using hand tools like snipes and shears as shown below.

Similar is the case with fruits like apple, plums, cherries, guava, pear, custard apple, jackfruit, water melon, litchis etc. Flowers grown in open field like roses, marigold, sunflower, tube rose, gladiolus, aster, etc which are marketed in bulk quantity are also cut by hand.

If observed keenly this activity of cutting the produce from the plant takes considerable amount of labour hours, it is tiresome and monotonous. Presently there are no mechanical means of carrying out this activity.

Thus taking into account the fact that all this activity is done in open field where abundant sunlight is available to tap solar energy the project concept proposes to design and develop a portable, light weight multi-purpose agriculture cutter and shear to carry the above mentioned activity using solar powered 12 Volt DC motor.

1.2 Energy used in Agriculture

As energy prices and volatility have increased in the past decade, incentives for energy efficiency and on-site renewable energy have emerged. Solar energy is emerging as a alternative energy source that ensures predictability, independence and even cost effectiveness for a number of agricultural applications. Potential is even higher for crop farming in India were in majority of agriculture work is done using farm labour using hand tools. Mechanization of farming activity will lead to reduced labour cost which has become extremely important in present age where shortage of farm labour due industrialization and inflation.

2. Literature Review

"Chiu-Fan Hsieh" in his paper titled "The effect on dynamics of using a new transmission design for eccentric speed reducers" has stated new transmission design for eccentric speed reducers that differs from that used with a traditional cycloid speed reducer. This paper proposes a new transmission design for eccentric speed reducers that differs from that used with a traditional cycloid speed reducer [1].

"R. Joshua, V. Vasu and P. Vincent" in his paper titled "Solar Sprayer - An Agriculture Implement" has stated that, Energy - demand is one the major thread for our country and finding solutions, to meet the "Energy - demand" is the great challenge [2].

"Ketchpel, 'Jr. et al." in his patent has stated that, A shear for grass and the like includes a tooth plate on which is pivotally mounted a plurality of shearing members, each including a narrow and resiliently deformable blade element [3].

"Hemant Ingale, N.N.Kasat" in his paper titled "Automated Solar Based Agriculture Pumping" he stated that, Solar power is absolutely perfect for use with irrigation systems for gardens, allotments, greenhouses, and polytonal. When the sun is shining you need more water and so the solar power is there for the pump. By adding a suitable deep-cycle leisure/marine battery, power can be made available 24 hours per day enabling watering in the evening - the best time to water plants in the summer so that the water has a chance to soak into the ground [4].

3. Design and Development

The development of concept is divided into three steps:

- Design Development, kinematic and strength analysis of multi-purpose agriculture cutter.
- Design Development kinematic and strength analysis of a shaft mounted speed reducer.
- Selection of proper solar panel to fulfil the requirement.

3.1 Design of Multi-purpose Agricultural Cutter

Design of multi-purpose agriculture cutter and shear suitable to cut all the above mentioned vegetables, fruits and flowers. It is obvious that there is a variety in stem diameter sizes and strength , thus a cutter with appropriate 'admit between shear blades' and appropriate strength is to be developed to achieve a cutting rate of minimum of 10 to 12 cuts per minute to achieve higher productivity and optimal use of labour by reducing labour fatigue and time. The concept of the shear will be developed on the basis of sketch shown below.

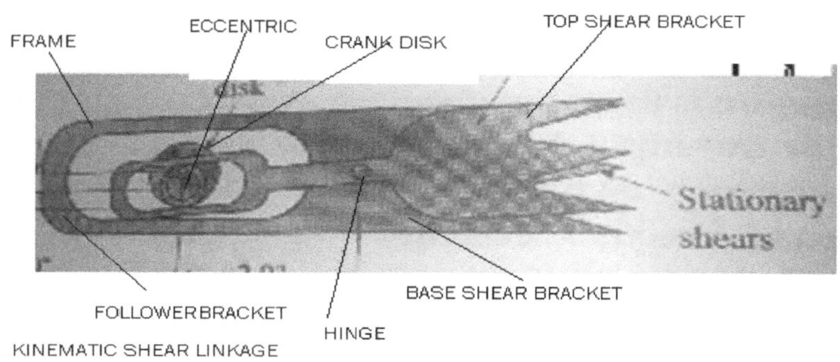

Figure 2: Agricultural Cutter

The kinematic linkage for the multi-purpose agricultural cutter cun shear mechanism is as above. The motor drives the eccentric shaft to rotate which will make the movable arm bracket to oscillate. The movable arm bracket then imparts this oscillating motion to the moving blade arms which will oscillate about respective hinges to bring into effect the cutting or shearing action between the shear blades.

3.2 Design of Shaft mounted speed reducer

Design of a shaft mounted speed reducer to power the shear mechanism with suitable shearing torque so that the stem diameters upto 8 to 10mm of various agricutlture produce . The arranegement of shaft mounted speed reducer is so selected that it gives maximum reduction ration (1:18) in minimum space and minimum weight owing to the fact that the agriculture cutter or shear is to be of minimum weight for effcient use the general arrangement of the gear box is to be as shown below.

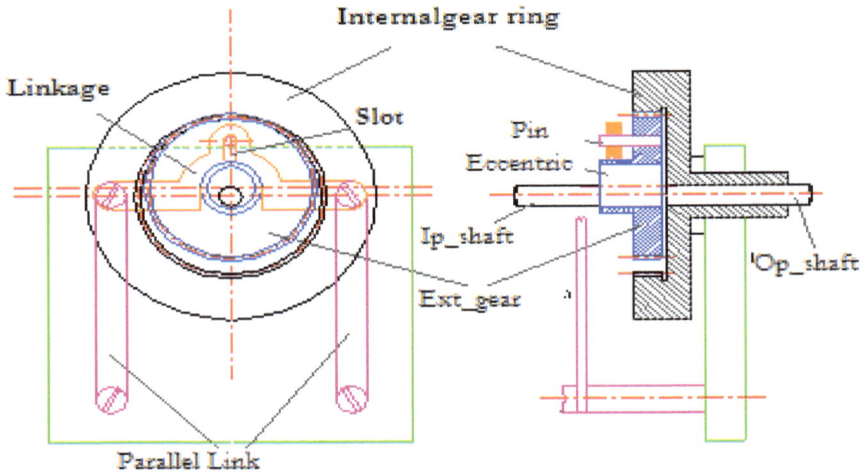

Figure 3: Speed Reducer

A standard internal gear and External gear are meshed without tooth interference. On the driving shaft is mounted an eccentric, the axis of the driving external gear follows the motion of eccentric, but is kept from rotating about its own axis by pin, which works in the slot. Parallel Linkage is actuated by the eccentric, which constantly maintains slot in a perpendicular position through the action of parallel links, pivoted on studs. Since the axis of external gear follows the motion of Eccentric and the gear does not rotate about its own axis, the motion imparted to the driven internal gear will be uniform.

➢ ***Formula***

$$\text{Output speed} = \text{Input speed} \times \left(\frac{N-n}{N}\right)$$

Where;

N = No. of teeth on internal gear

n = No. of teeth on external gear

4. Motor Selection

Thus selecting a motor of the following specifications

12 V DC motor

Power = 1/15hp=50 watt

Speed= 0-2000 rpm (variable)

Motor is a 12 V DC motor, Power 50 watt, the speed of motor is varied by means of an electronic speed variator. Motor is a commutated motor i.e. the current to motor is supplied to motor by means of carbon brushes. The power input to motor is varied by changing the current supply to these brushes by the electronic speed variator; thereby the speed is also is changes. Motor is foot mounted and is bolted to the motor base plate welded to the base frame of the indexer table.

5. Results and Discussion

Testing of the agriculture cutter cum shear mechanism to determine

1. Maximum admit between blade capacities of cut.
2. Maximum number of cuts per minute for various item sizes.
3. Comparison of the newly developed agriculture cutter cum shear mechanism with conventional practices on grounds of productivity improvement, savings in labour cost and payback time of mechanism.

References

[1] Chiu-Fan Hsieh, "The effect on dynamics of using a new transmission design for eccentric speed reducers", Mechanism and Machine Theory 80 (2014) 1–16.

[2] R. Joshua, V. Vasu and P. Vincent, "Solar Sprayer - An Agriculture Implement", International Journal of Sustainable Agriculture 2 (1): 16-19, 2010.

[3] Hemant Ingale, N.N.Kasat, "Automated Solar Based Agriculture Pumping" International Journal of Advanced Research in Computer Science and Software Engineering 2 (11), November- 2012, pp. 407-410.

[4] J.G. Blanche, D.C.H. Yang, "Cycloid drives with machining tolerances", ASME J. Mech. Transm. Autom. Des. 111 (1989) 337–344.

[5] D.C.H. Yang, J.G. Blanche, "Design and application guidelines for cycloid drives with machining tolerances", Mech. Mach. Theory 25 (5) (1990) 487–501.

[6] J.B. Shung, G.R. Pennock, "Geometry for trochoidal-type machines with conjugate envelopes", Mech. Mach. Theory 29 (1) (1994) 25–42.

[7] F.L. Litvin, P.H. Feng, "Computerized design and generation of cycloidal gearings", Mech. Mach. Theory 31 (7) (1996) 891–911.

Experimental Study on Performance Characteristics of Single Cylinder Four Stroke Diesel Engine Using Blends of Diesel & Palm Oil Methyl Ester as an Alternate Fuel

Niraj Raja[1] , Dr. M. Basavaraj [2] , Prof. Avinash Khanderao[3]

[1]*M.Tech, Heat Power Engineering, Department of Mechanical Engg., B.I.T. Ballarpur, Dist. Chandrapur, Maharashtra, India* niraj28raja@rediffmail.com
[2]*Professor, Department of Mechanical Engg, B.I.T. Ballarpur, Dist. Chandrapur, Maharashtra, India.*
[3]*Asst. Professor, Department of Mechanical Engg, B.I.T. Ballarpur, Dist. Chandrapur, Maharashtra, India.*

Abstract

Petroleum based fuels play a vital role in rapid depletion of conventional energy sources along with increasing demand and also major contributors of air pollutants. Major portion of today's energy demand in India is being met with fossil fuels. Hence it is high time that alternate fuels for engines should be derived from indigenous sources. As India is an agricultural country, there is a wide scope for the production of vegetable oils (both edible and non-edible) from different oil seeds.

The present work focused only on vegetable oils as fuel for engines, as the edible oils are in great demand and far too expensive. Experiments are carried out in a more popular petter type single cylinder, water cooled engine. Major problems associated with vegetable oils are higher viscosities, lower heating values, raise in stoichiometric fuel air ratio and thermal cracking. The focused is mainly on the utilization of Palm oil methyl ester and their blend with diesel in diesel engine. The neat oil blends with diesel were heated before entering into combustion chamber. The heating value depends on the increase in percentage of neat oils in mixture to reduce the viscosity of the fuel. The performance parameters of the test engine Viz. Brake thermal efficiency is decreased, Brake specific fuel consumption and Exhaust gas temperature are increased for all neat oils compared to diesel. From the experimentation, it is observed that 25% of neat oil mixed with 75% of diesel is the best suited blend, without heating and without any modification of the engine.

Keywords: alternate fuels, diesel engine, palm oil, performance parameters.

1. Introduction

India is one of the fastest developing countries with a stable economic growth, which multiplies the demand for transportation in many folds. Fuel consumption is directly proportionate to this demand. India depends mainly on imported fuels due to lack of fossil fuel reserves and it has a great impact on economy. India has to look for an alternative to sustain the growth rate. Bio-diesel is a promising alternative for our Diesel needs. With vast vegetation and land availability, certainly bio-diesel is a viable source of fuel for Indian conditions. Recent studies and research have made it possible to extract bio-diesel at economical costs and quantities. The blend of Biodiesel with fossil diesel has many benefits like reduction in emissions, increase in efficiency of engine, higher Cetane rating, lower engine wear, low fuel consumption, reduction in oil consumption etc. It can be seen that the efficiency of the engine increases by the utilization of Bio-diesel. This will have a great impact on Indian economy.

History of vegetable oils:
India is importing crude petroleum & petroleum products from Gulf countries. Indian scientists searched for an alternate to diesel fuel to preserve global environment and to withstand economical crisis. So, vegetable oils from plants both edible, crude non-edible and Methyl esters (Bio-diesels) are used as alternate source for Diesel oil. Bio-diesel was found as the best alternate fuel, technically and environmentally acceptable, economically competitive and easily available. In the view of future energy crisis, the author has chosen Methyl esters of Palm oil for investigation in diesel engine. In this the author has focused on Methyl esters of Palm oil & its blends with diesel have been tested as alternate fuel for C.I. Engine.
The important physical and chemical properties of the above said oil and their blends are determined. The effect of blending this oil with diesel was studied. Effect of temperature on viscosity has been studied. Then the suitability of vegetable oils as alternate fuel is identified through the evaluation based on the performance parameters. These results are compared to those of diesel.

2.1 PROPERTIES OF VEGETABLE OILS USED IN TEST ENGINE

Palm oil has pleasant odour and taste. It is stable and resistant to rancidity. The colour of palm oil varies from yellow to deep orange. Inter esterification of palm oil produces two fractions. Palm oil obtained at low melting point called "Olein" and the oil obtained at high melting point called "Stearin". Oil palm fruits are oval-shaped sessile drupes. Palm oil contains some triglyceride species, which are completely saturated. The iodine value of palm oil is lower (44-58) when compared to other vegetable oils because of high proportion of saturated fatty acids. Palm oil is solid at ambient temperature and fluid in tropical and subtropical climates with certain fractions held in crystalline form. It is used in manufacturing plastics, fibers and soaps. It is available in Asia, Africa, Indonesia, Nigeria and Malaysia.

Property of oils	Neat Diesel (B00)	B25	B50	B75	Methyl Ester of Palm Oil (B100)
Density(gm/cc)at 40^0C	830	842	851	858	876
Viscosity(cst)	2.7	3.2	3.91	4.78	5.71
Flash point(0C)	64	NA	NA	NA	174
Fire point(0C)	69	NA	NA	NA	179
Calorific values(KJ/Kg)	42500	41480	40100	39200	38500
Cetane number	51	NA	NA	NA	51

2.2 TRANSESTERIFICATION PROCESS

The transesterification process will be adopted for the preparation of ethyl ester or methyl ester of vegetable oil. In the preparation of ethyl ester (biodiesel), five distinct stages will be involved.

1) Heating of oil.
2) Preparation of alkaline mixture.
3) Adding of alkaline alcohol to oil and stirring the mixture.
4) Settling of separation of glycerol.
5) Washing of methyl ester with water.

The biodiesel can be obtained by transesterification of vegetable oil using either ethanol or methanol as the Trans-esterifiction agent. Transesterification reaction is a stage of converting oil or fat into methyl ester or ethyl esters of fatty acids which constitutes to biodiesel. Biodiesel (methyl esters) is obtained though the reaction of triglycerides of vegetable oils with an active intermediary formed by reaction of an alcohol with a catalyst. The general reaction for obtaining biodiesel though transesterification.

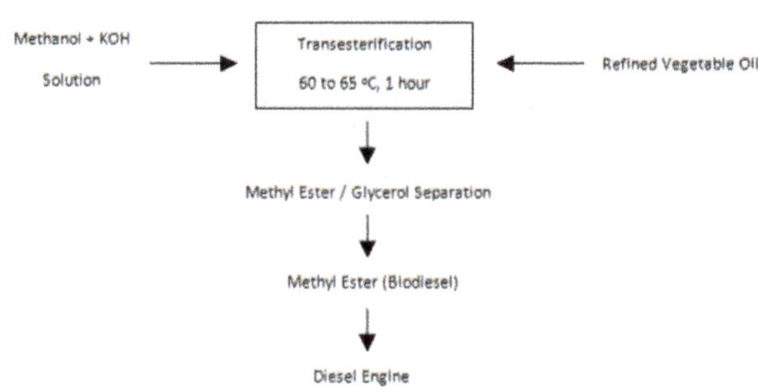

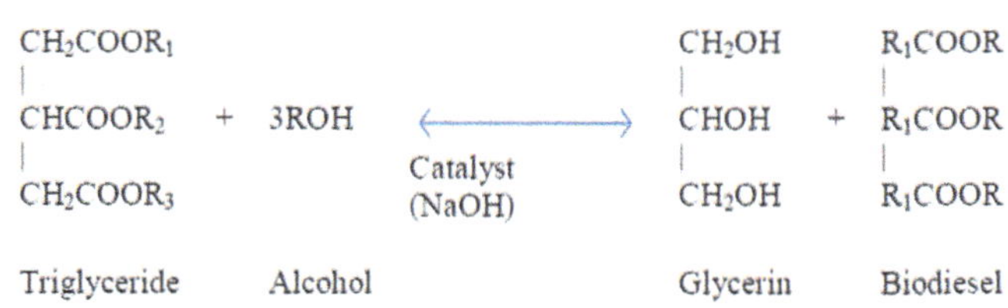

Transesterification reaction

Figure 1. Procedure of manufacturing biodiesel through transesterification with vegetable oil

The reaction requires heat and a strong catalyst (alkalis, acids, or enzymes) to achieve complete conversion of the vegetable oil into the separated esters and glycerin. Generally potassium hydroxide (KOH) and sodium hydroxide (NaOH) are used as alkaline catalysts with methanol (CH_3OH) for production of biodiesel. The alcohol & catalyst is charged into a closed reaction vessel and the vegetable oil is added. The reaction is heated to the boiling temperature of the alcohol (normally 50-60 deg C) and is refluxed for a certain length of time under agitation. When agitation is stopped (after 60minutes) the reaction mixture separated into an upper layer of methyl ester and lower layer of glycerol diluted with un reacted methanol. Once the reaction is complete, two major products are produced: Glycerin and biodiesel (methyl ester). The quantity of glycerin varies according varies according to the vegetable oil used, the process used, and the amount of the excess alcohol that is used in the reaction. The fatty ester produced in the upper layer is neutralized and vacuum distilled for the removal of excess methanol. The methyl ester produced from the reaction is then washed with hot water and separated out by centrifugation

2.2 EXPERIMENTAL WORK

The Engine chosen to carry out experimentation is a single cylinder, four stroke, vertical, water cooled, direct injection diesel engine.

Single Cylinder Diesel Engine Test Rig With Rope Brake Dynamometer. The J.P.T.I.' unit consist of a 4-stroke diesel engine coupled to a rope brake dynamometer. A water cooled brake drum along with spring balances is provided. A Rope brake arrangement is provided to load the engine.

SPECIFICATION:-

J.P. Techno Instru. Make A Single Cylinder Four Stroke Diesel Engine

Engine Power: 3.75kW 1500rpm 5hp	Max speed: 1500rpm	
Cylinder Bore: 84.5mm	Stroke Length: 110mm	
Displacement: 660cc	Compression Ratio: 17.5:1	
Speed: Constant speed type	Cooling: Water cooling type	Type of start: Crank shaft type starting

1) Diesel engine of 5 H. P. capacity ,1500 R. P. M. 4-stroke vertical, single cylinder engine.
2) Loading device - Rope brake consisting of a brake drum of size suitable diameter & rope brake arrangement with rope ,hand wheel, spring balances.
3) Fuel measuring arrangement-Consisting of a fuel tank, mounted on a strong iron stand, measurement of a fuel consumption by a burette & 3- way cock. Connecting tube, a stop watch.
4) Cooling water arrangement:- Cooling pipes, thermocouples, measuring flask to determine the discharge of cooling water & heat carried away by cooling water.
5) Exhaust gas temperature (digital temp. indicator)
6) Air intake measurement :- Air tank reservoir of size 0.5 m. x 0.5 m. x 0.5 m. with orifice plate, differential manometer, connecting hose.

The engine has a rope brake dynamometer to measure its output. The experiments are conducted for variable loads like 0, 25%, 50%, 75% 7 full load conditions at rated speed. The blends of diesel and methyl ester of palm oil are prepared with different proportions like 25% methyl ester of palm oil + 75% diesel and termed it as B25 (Biodiesel 25), similar to B25; B50, B75 and B100 (neat palm oil) are prepared and used in experimentation. The palm oil and their blends with diesel are heated externally to a required temperature as stated earlier before injecting into the test cylinder. All the observations recorded were replicated thrice to get a reasonable value. The performance parameters such as Brake Thermal Efficiency (ηB.Th.), Brake Specific Fuel Consumption (bsfc), Exhaust Gas Temperature (EGT). These performance parameters of oils are compared to those of pure diesel.

3. RESULTS & DISCUSSION

The experimental investigations are carried out using the above said oils and their blends on the test engine. The detailed analyses of these results are as discussed below.

Fig. 1 shows the variation of Brake Thermal efficiency with Brake power output for Methyl Ester of Palm oil and its blends with Diesel in the test engine. For 25% blend of Palm Stearin slightly lower Brake thermal efficiency compared to diesel. Neat palm stearin oil has lower Brake thermal efficiency at all loads.

Fig. 2 shows the variation of Brake specific fuel consumption with Brake power output for Palm Stearin oil and its blends with Diesel in the test engine. Neat Palm Stearin oil has higher bsfc compared to diesel. 25% blend of Palm Stearin oil has lower bsfc compared to all other blends.

Fig. 3 shows the variation of Exhaust Gas temperature with Brake power output for Palm oil and its blends with diesel in the test engine. 25% blend of Palm stearin has lower EGT compared to all other blends for all loads.

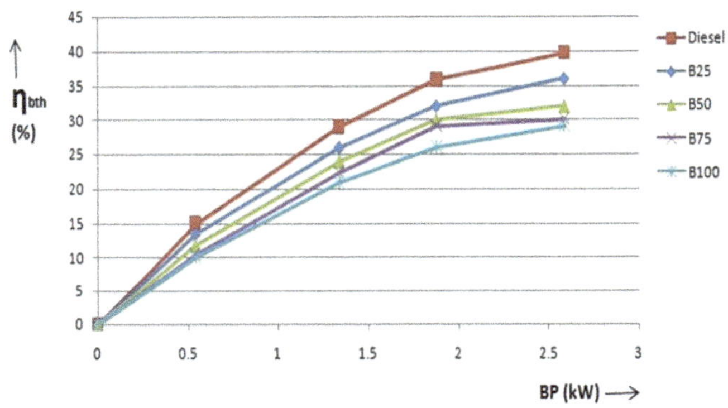

Fig. 1 Break Power (BP) Vs Break Thermal Efficiency (η_{bth})

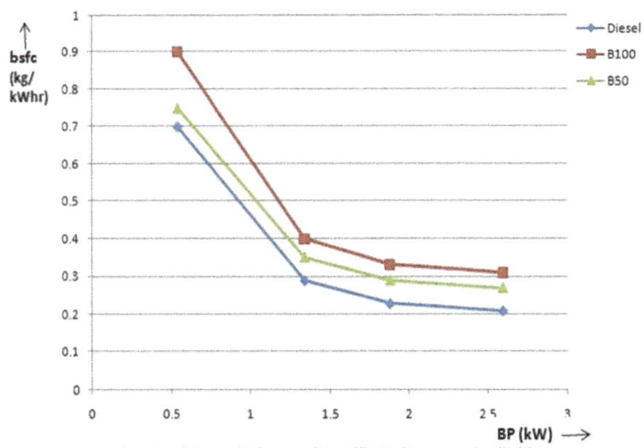

Fig. 2 Break Power (BP) Vs Break Specific Fuel Consumption (bsfc)

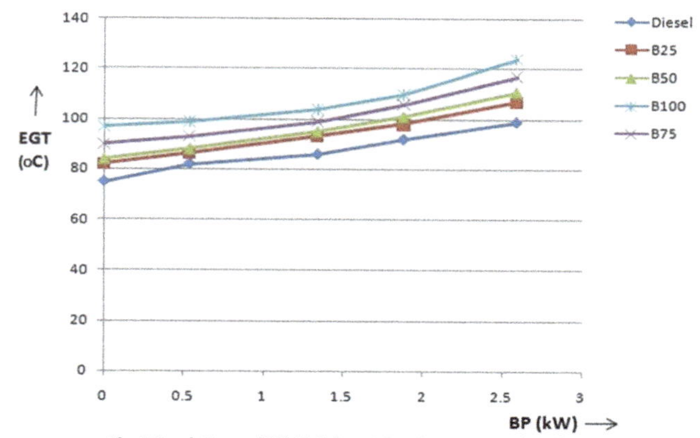

Fig. 3 Break Power (BP) Vs Exhaust Gas Temperature (EGT)

4. CONCLUSION

Performance parameters of engine such as Brake thermal efficiency, Volumetric efficiency are decreased, Brake Considering the need for alternate fuels, the experimental investigations are carried out in the present work in order to run the existing diesel engines with non-edible vegetable oils. specific fuel consumption and Exhaust gas temperature are increased for all neat oils and their blends compared to those of diesel. This is because of high viscosity coupled with lower heating value of the fuels. Physical and chemical properties of the above mentioned oil blends were determined. The performance parameters of chosen neat oil and their blends were evaluated. These results are compared to those of diesel.

5. SUMMARY OF RESEARCH

1. The physical and chemical properties of chosen non edible palm oil and their blends with diesel are determined and analyzed.
2. The analysis for performance characteristics such as brake thermal efficiency, brake specific fuel consumption, exhaust gas temperature, volumetric efficiency of chosen oil and their blends with diesel are done through graphs.
3. The performance characteristics of five chosen oils are compared to the other non-edible oils already available in the literature for validation.
4. The suitability of oils as alternate fuel in C.I. Engines is checked through validation process.

6. FUTURE ISSUES

Investigations are to be carried out on different blends of esters with diesel to determine better performing blends. Investigations have to be carried out on combustion characteristics. Investigations are to be carried out on emission characteristics by using multi cylinder engine.

7. DISCLOSURE STATEMENT

There is no special financial support for this research work from the funding agency.

8. ACNOWLEDGMENT

Im highly obliged to express my deep sense of gratitude and grateful to thanks my Project Guide Dr. M. Basavaraj & Co- Guide Prof. Shiekh Yasin for their valuable guidance and support which led to successful and timely completion of our report. Last but not the least Im thankful to Miss. Supriya B. A Chemical Research analyst IBDC Baramati for her valuable time and technical knowledge which helped me in completion of my research work.

REFERENCES

[1] B. Deepanraj, C. Dhanesh "Use of palm oil biodiesel blends as a fuel for compression ignition engine" in American Journal of Applied Sciences 8(11) pp1154-1158, 2011 ISSN 1546-9239

[2] S.S. Wiraman, A.H. Tambunan, " The effect of palm biodiesel fuel on the Performance and Emission of the automotive diesel engine" Agricultural Engg. International, the CIGR Ejournal, Manuscript EE 07 005 Vol. X April 2008

[3] H. Mulimani M.C. Navindgi, " Extraction of biodiesel from vegetable oils and their comparisons" Issue 2, Volume 2 International Journal of Advanced Scientific Research and Technology. ISSN 2249-9954

[4] C. Solaimuthu, D. Senthilkumar , " An experimental investigation of performance, combustion and emission characteristics of mahua oil methyl ester on four stroke direct injection diesel engine" in Indian Journal of Engg. & Material Sciences. Vol 20, February 2013, pp 42-50

[5] C. Naga Prasad, Vijaya Kumar Reddy, "Performance & Emission characteristics of a diesel engine with castor oil" in Indian Journal of science & technology" Vol. 2 No. 10 (Oct. 2009) ISSN: 0974-6846

[6] K. Arumugam, S. Veeraraja, " Methyl ester of vegetable oil – A promising alternate fuel" in International Journal of Applied Engg. Research Vol. 8, No. 15, 2013 pp. 1765-1772

[7] Kalam MA, Masjuki HH .Biodiesel from palmoil an analysis of its properties and potential. BiomassBioenergy 2002; 23:471–9.

[8] Anand K, Sharma RP, Mehta PS. Experimental investigationson combustion ofJatropham methyl ester in a turbochargeddirect-injection diesel engine. P MechEng D-JAut 2008; 2221865e77.

[9] A. Haiter Lenin, R. Ravi, "Performance Characterstics of a diesel engine using Mahua Biodiesel as alternate fuel" in Iranica Journal of Energy & Environment 4(2): 136-141, 2013, ISSN 2079-2115

17

DESIGN AND ANALYSIS OF AN ATMOSPHERIC VORTEX ENGINE

[1]Rathnavel S, [2]Vijaya sankaran K

[1] *PG Scholar, Department of Aeronautical Engineering, Regional Centre of Anna University, Tirunelveli-627007, India. (rathnavelaero@gmail.com)*

[2] *PG Scholar, Department Aeronautical Engineering, Park College of Engineering and Technology, Coimbatore-641668, India. (vijayasankaran@live.com)*

Author Correspondence: *Vijaya sankaran K, 2/132-1 modern nagar, collectorate post, virudhunagar-626002*

ABSTRACT

The demand for power has been increased nowadays in many folds and the need for a better, reliable and cleaner source of energy has become essential. Ultimately our aim is to fulfill the requirements of power demand. By research it is found that atmospheric vortex engine (AVE), the twin of solar chimney, seems to be a promising source and has many advantages than the latter. Through this research it is possible generating power from the residual heat source from the industries/thermal power plants. The AVE model is designed using computerized modeling and analyzed it using computational fluid dynamics. By varying the parameters, that influence the performance of AVE, it is studied to obtain an optimum design and operating conditions. Also the AVE scaled model is fabricated and passed the flow to study about the vortex practically. It is found that it has enough strength to rotate a small turbine placed inside it. Also the smoke has been passed and small particles to visualize the flow. In this research the results of the CFD analysis to show that AVE is capable of producing a sustained vortex, to generate power. This paper also serves as a base for future research.

Keywords: AVE, Alternate power, Updraft, Vortex.

1 . INTRODUCTION

In this vast growing world, need for a clean energy beacons among the researchers from around the world. Some of the energy sources have been identified as renewable, most of them from the sun or movements in air or water currents due to earth's rotation. Each and every method has its own advantages as well as limitations. Solar panels involve the usage of rare elements and are costly, various other methods such as geo thermal, tidal etc., require complex assemblies and are not reliable throughout a year. The concept of updraft and vortex is used to produce power from the source. Sun's heat causes changes in density and it produces updrafts [2].

Power from the tornadoes is possible but it was not obtained in the real case [1]. The atmospheric vortex engine has a combination of principle of tornadoes and the principle of solar chimney [8, 9]. The free vortex is the vortex circulates faster than the fluid far from the centre than in the centre [5]. The forced vortex is the fluid rotates as a solid body and this is used to create the vortices. The properties of the vortex are fluid pressure is lowest in the centre and rises from the centre also it starts and ends only in the boundary of the fluid or form closed loops [1, 3-5, 15]. The change in temperature causes change in density so the movement of lighter air to fill vacuum is termed as Updraft [8-10].

2. BASIC PRINCIPLE

The solar chimney is a base idea for AVE and the principle behind solar chimney is the air is heated by solar radiation under a transparent roof the roof and the natural ground below form an air collector and in the middle of the roof is a vertical tower as hot air is lighter than cold air it rises up the tower. The solar chimney converts solar energy into thermal energy and then to kinetic energy and then to electrical power. The atmospheric vortex engine (AVE) uses an artificially created vortex to capture the mechanical energy produced during upward heat convection. The heat source can be solar energy, warm seawater or waste industrial heat. The AVE has the same thermodynamic basis as a solar chimney.

Thus solar radiation or heated air causes a constant updraft in the tower and the energy contained in the updraft is converted into mechanical energy by pressure-staged turbines at the base of the tower, and into electrical energy by conventional generators.

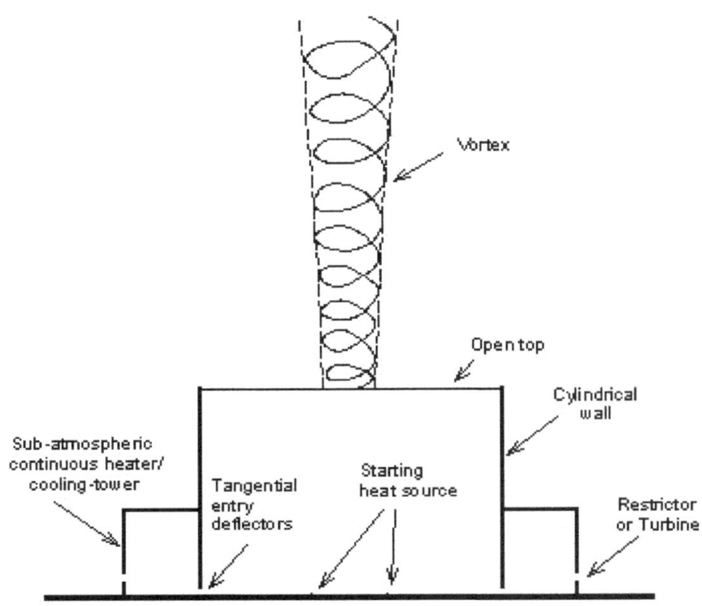

FIG 1: SCHEMATIC REPRESENTATION OF AVE

In a forced vortex the fluid rotates as a solid body. The fluid has vorticity of 2ω everywhere, and the free surface (if present) is a paraboloid. The tangential velocity is given by

$$v_\theta = \omega r \tag{1}$$

Thus from the knowledge given about vortices we can observe, in forced vortex the fluid has a greater velocity at the outer surface and gradually decreases towards the centre.

3. METHODOLOGY

In this research Gambit is used for designing the vortex engine and fluent for analyzing the model for flow control.

To study the vortex flow in real case the model to produce the vortex has been created. If a turbine is placed inside the model and it is capable to rotate by the power of vortex. The vortex flow can be visualized by the smoke test.

4. MODELLING

The model is designed using gambit by the following specifications. The specifications used are studied from the various papers which are using the concept of solar chimney. The real case chimney is not possible for analysis, scaled down dimensions are used. It is first developed by Louis M.Michaud to derive the power from the updraft and vortex creation. Meshing of the model is done in Gambit and tetrahedral Mesh is used to mesh the model and it is good for solid model.

Parameter Name	Parameter Description	Base Value (mm)
d1	Inner octagonal cylinder diameter	1000
d2	Outer octagonal cylinder diameter	1050
d3	Roof opening diameter	300
d4	Floor opening diameter	0
g1	Gap between deflector	50
h1	Tangential entry height	200
h2	Octagonal cylinder height	200
h3	Upper cylinder height	0
Z	Domain height	2000

TABLE 1: Dimensional specifications for prototype model

The model is created as in Fig. 1 and it is meshed in gambit as in Fig. 2.

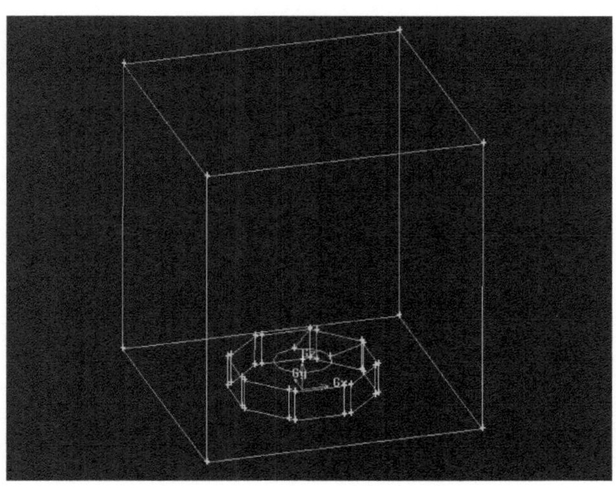

Fig. 1 Model in Gambit

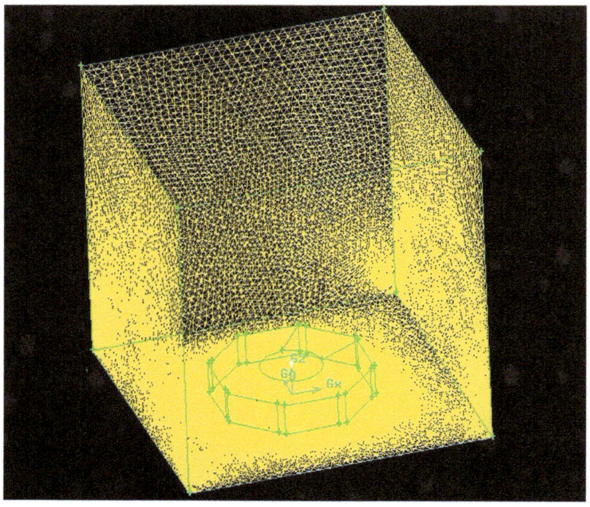

Fig. 2 Meshed Model

5. ANALYSIS

The model is analyzed in the fluent. By setting up the boundary conditions it can analyze the model. For turbulence model in Computational fluid dynamics k-ε model is used. The pressure and temperature value for the boundary condition is also necessary. We have applied the pressure value for AVE inlet, AVE top and Domain top. Similarly for the temperature for the boundary conditions are applied for AVE inlet, domain wall and domain top. After the boundary condition has been setup we have analyzed using fluent solver and obtained the result.

Boundary name	**Boundary condition**
Cubic domain side wall (4)	Pressure Inlet (Atm Pr, Ambient Temp T)
Cubic domain roof	Pressure outlet (Atm Pr, Ambient Temp T)
Cubic domain base	Free-slip wall
Tangential air inlet (8)	Pressure inlet (Atm Pr, Ambient Temp $T + \Delta T$)
AVE base	No-slip wall (Temp $T + \Delta T$)
AVE roof opening	Pressure outlet (Atm Pr, ambient temp $T + \Delta T$)
AVE octagonal cylindrical wall	No slip wall
AVE roof	No slip wall

TABLE 2: Boundary conditions for AVE in Fluent

The plots of velocity magnitude, temperature and velocity vector distribution is discussed in the results and discussion

6. FLOW VISUALIZATION

The model is designed in wood and the smoke is also passed to visualize the vortex. To design this model the scaled down dimensions of AVE is used. It is octagonal in shape and it has eight inlets from each side of the octagon.

Floor opening of the AVE is fully closed and it has roof opening. The movable guide vanes like is deflector is placed for each inlet. The vanes can be opened or closed. By using the blower the source of air is passed to the experimental model of AVE. The passing of air from blower also produces heat air and it creates some pressure difference. The pressure difference produces the vortices and it will rotate the turbine if it is placed on the floor of AVE. It is due to the production of vortex. The speed of the turbine is also measured by varying the roof opening diameter. The view of vanes opened and closed position is shown in the Fig. 3

Fig. 3 AVE with vanes opened

To visualize the vortex the smoke testing method is used. The air is passed in the inlet and the vortex strength is observed.

Fig. 4 Visualization of vortex by smoke

7. RESULTS AND DISCUSSION

. The contour plots of velocity magnitude, static temperature, tangential velocity in k-ε model

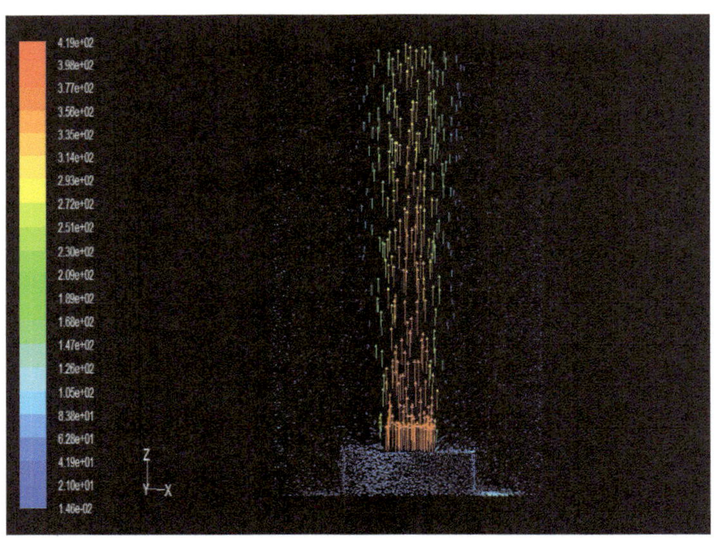

Fig. 5 Velocity Vector for AVE

Fig 5 illustrates that there is a natural updraft. Then in the velocity contour, velocity is varying linearly.

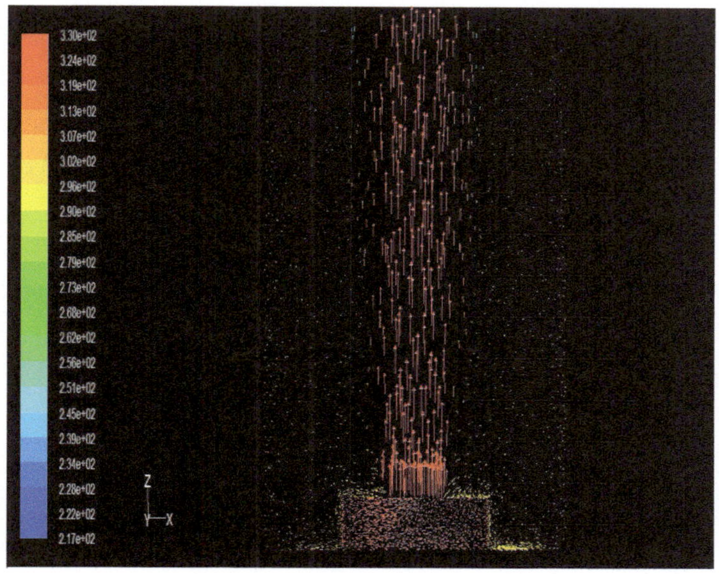

Fig. 6 Temperature Contour for AVE

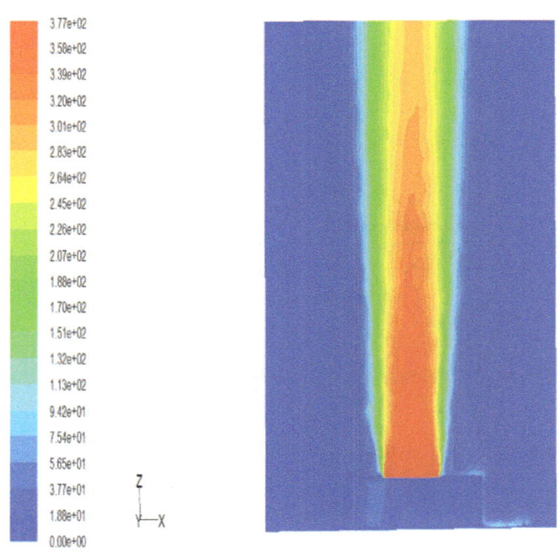

Fig. 7 Velocity Contour for AVE

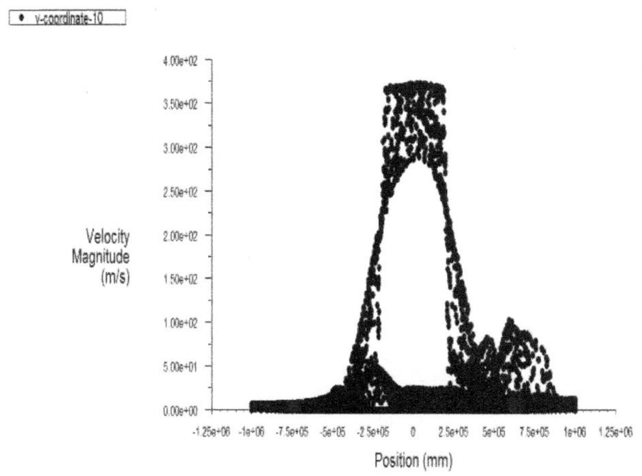

Fig. 8 Velocity Magnitude Vs position of AVE

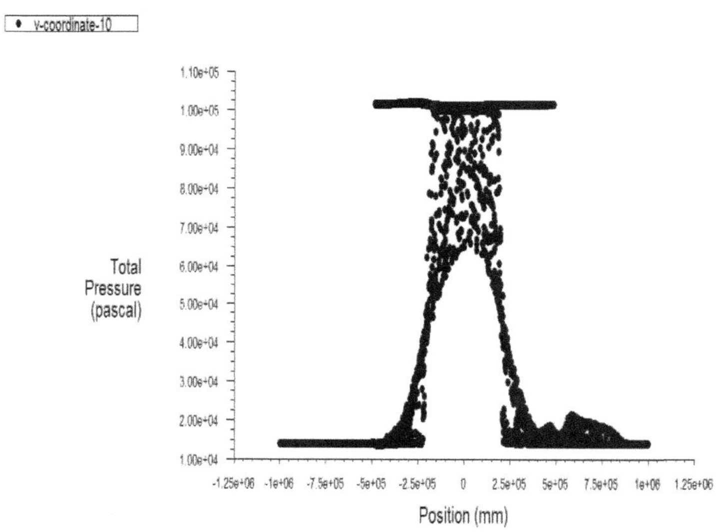

Fig.9 Total Pressure Vs position for AVE

Then from the temperature contour it is that some convective heat transfer is indicated by upward moving arrows.

From the above results it is clearly shown that the turbine can be rotated by the velocity and also the vortex is created in the flow which rotates the turbine.

The graphs are also showing good results that the updraft has been created.

8 . CONCLUSION

From this article, it is found AVE design has capability of producing vortex. It can be applied for large scale by this method. From the computational results one can assure that the feasibility of AVE in real case is possible and it is no more conceptual.

The advantage of AVE is eco-friendly and also there is no carbon emission. It can use industrial effluents and non hazardous. It is the combination of solar and wind turbine so it has more efficient than other power generating devices. Some disadvantages of the AVE are the height of tower and some noise is produced in the tower. Now it cannot be used on densely populated areas and it requires some additional start-up energy. In future the operating difficulties of AVE in many conditions will be analyzed and it can be used for many purposes.

REFERENCES

[1] Church, C.R, Snow, J. T, Baker, G. L, Agee, E.M, "Characteristics of tornado like vortices as a function of swirl ratio", A laboratory investigation, Journal of the Atmospheric Sciences ,1979.

[2] Djezzar mahfoud, Tayebi tahar, "Numerical simulation of natural convection in a solar chimney", 2003.

[3] Diwakar Natarajan, "Numerical Simulation of Tornado Like vortices", University of western Ontario, Canada, 2011.

[4] Fujita T, "Tornadoes and downbursts in the context of generalized panetary scales", Journal of the Atmospheric Sciences, 1981.

[5] Grasso L. D, and Cotton W. R, "Numerical simulation of a tornado vortex", Journal of the Atmospheric Sciences, 1995.

[6] Haaf W, "Solar Chimneys -Part II: Preliminary Test Results from the Manzanares Pilot Plant", International Journal of Solar Energy, 1984

[7] Hangan H, and Kim J, "Swirl ratio effects on tornado vortices in relation to the Fujita scale", Journal of Wind and Structures, 2008.

[8] Michaud L. M, "On the energy and control of atmospheric vortices", Journal de Recherches Atmospheriques 1977.

[9] Louis M Michaud p Eng "The Atmospheric Vortex Engine", AVEtec Energy Corporation, 2008.

[10] Louis M Michaud "Unrestrained Expansion- A source of Entropy", Submitted to AGU Fall meeting, 2005.

[11] N Renno, "A thermodynamically general theory for convective vortices", Tell us A, Vol 60-4, pp. 688-699, 2008.

[13] L Michaud, "Vortex process for capturing mechanical energy during upward heat-convection in the atmosphere", Applied Energy, Vol62 (4), pp141-161, 1999.

[14] Christopher R. Church, Karen A. Kosiba, J. David Cleland, and Christopher P. Beer, "The Formation and intensification of supercritical tornado like vortices – a laboratory study", Miami University, 2009.

AUTHOR BIOGRAPHY

1. **Rathnavel S** is pursuing Master Degree in Aeronautical Engineering from Regional Centre of Anna University, Tirunelveli. He obtained Bachelor Degree in Aeronautical Engineering from Anna University; Chennai, India (2013).His area of interest is Aerodynamics and Computational Fluid Dynamics

2. **Vijaya sankaran K** is pursuing Master Degree in Aeronautical Engineering from Park college of Engineering and Technology, Coimbatore. He obtained Bachelor Degree in Aeronautical Engineering from Anna University; Chennai, India (2013).His area of interest is Aerodynamics, Flight mechanics and Aircraft Design.

REVIEW ON E-BRAKE SYSTEM FOR CABLE TROLLEY SYSTEM

Tushar S. More[1], Sharayu U. Ratnaparkhi[2]
[1]PG Student, tushar.more33@gmail.com
[2]Assistant Professor, shrayu.ratnaparkhi@gmail.com
Author Correspondence: Sai Shilpa society, Bibvewadi, Pune.

Abstract

The goal of this project is to provide an emergency braking system which will stop the descent of the trolley automatically and allow for a manual, controlled descent to be performed by an occupant of the trolley if needed. The braking system must be reliable, be able to operate after long period of non-use and operate without destruction of major braking system components when used. The concept in which a brake calliper applies braking force directly to a secondary cable was selected. A provision for manual operation of the calliper is added to the system.

Keywords: Vertical motion control, Re-levelling control, Brake calliper, Brake lining system, Brake pad evaluation.

1. Introduction

The trolley was intended to haul people and cargo up the steep slope to the building. The trolley operates on a single pull cable and at the inception of this project had no means of controlling the trolley should the pull cable break or the winch malfunction. An emergency braking system is required to keep people and cargo safe in the event of a cable or winch malfunction. The goal of this project is to provide an emergency braking system for the trolley that will activate automatically in the event of a cable or winch malfunction. The braking system will stop the descent of the trolley automatically and allow for a manual, controlled descent to be performed by an occupant of the trolley if needed. The braking system must be reliable, be able to operate after long periods of non-use, and operate without destruction of major braking system components when used.

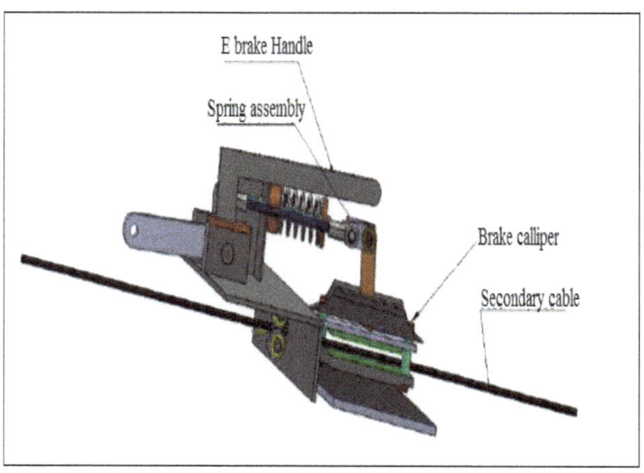

Figure 1: Design of E-braking system

1.1 Problem Statement

An emergency braking system for a funicular trolley located is to be designed and prototyped. The braking system is to operate automatically in the event of a pull cable or winch gearbox malfunction. The braking system should either stop the trolley, or allow the trolley to descend to the bottom of the way at a controlled velocity of no more than 2 ft/sec. If the braking system is to stop the trolley, a provision for manual operation of the brake may be included to allow an occupant on the trolley to descend to the bottom of the trolley way. The braking system must be reliable, mechanical (non-electrical), and operate without the destruction of major braking system components. Documentation of the mechanism, including Bill of Materials (BOM), operating and maintenance instructions (O&M), and Assembly Drawings are to be delivered with the prototype.

1.2 Concept and Ideation

The product utilizes a secondary cable stretched from the top to the bottom of the trolley way. The secondary cable passes through a mechanical brake calliper which is attached to the trolley. A linkage set connects the winch pull cable to the calliper, and a spring is utilized to apply braking force to the calliper in the event of an emergency. Fig 1 provides models of the complete braking system. The braking system features three distinct subunits, the calliper assembly, the calliper activation assembly, and the cable guides.

The internal search process produced feasible concepts. These concepts fall into 3 basic categories: direct cable friction systems, brakes on a solid member, and brakes controlling the rotational motion of a shaft with rotation caused by an independent cable looped around a pulley connected to the shaft.

The concept in which a brake calliper applies braking force directly to a secondary cable was selected. A provision for manual operation of the calliper by an occupant of the trolley was added to the system.

2. Literature Review

Sung-Min Moon , Jaemyung Huh , Daehie Hong, Seunghoon Lee c, Chang-Soo Hanc, in his paper, "Vertical motion control of building Façade maintenance robot with built-in guide rail (2014)" states that, " Many researches on an automatic building façade maintenance system have been conducted to satisfy the increasing demands of façade maintenance. [1].

"Daniel blaire boren, Brett j Epstein, c. Hass, Hickman in his paper, "United States Patent (May 29,2001)" states that , "An apparatus for transporting a rider between two points on a Cable having a plurality of Supports Configured to suspend the cable or rope between two points of unequal elevation and a trolley [2].
Otterson et al., Gary Erickson, Smarte Trolleye, in his paper, "United States Patent (May 29, 2001)" [3] states that," A nestable trolley, of the type guided by a walking attendant, for conveying a material such as luggage. The trolley has a frame and is supported by two rear wheels and a front rotatable caster. The brake is automatically engaged when the trolley is left. [3].

Tretsiak, Dzmitry, Kliauzovich, Siarhei, in his paper, "Belarusian National Technical University, Belarus (2006)",. State that "The current tendencies in automotive industry need intensive investigation in problems of interaction of active safety systems with brake system equipments. [6].

3. Design Requirement

The design team determined that the following requirements were highly important to the success of the concept:

- Brake system will automatically stop the trolley (total combined weight < 800 lbs) in an emergency
- The system should have a manual brake release on the trolley
- Stopping acceleration will not be greater than the acceleration experienced at start up
- All components must be mechanical
- The braking system should capture the trolley to the track while still allowing for the trolley to be removed easily when necessary
- The minimum factor of safety for all brake system components must be at least 2

4. Design and Development

Design consists of application of scientific principles, technical information and imagination for development of new or improvised machine or mechanism to perform a specific function with maximum economy and efficiency. Hence careful design approach has to be adopted. The total design work has been split up into two parts.

- Design of System
- Mechanical Design

4.1 Design of System

System design mainly concerns with various physical constrains, deciding basic working principle, space requirements, arrangements of various components etc.
Following parameters are looked upon in system design.

- Selection of system based on physical constraints. The mechanical design has direct norms with the system design hence system is designed such that distinctions and dimensions thus obtained in mechanical design can be well fitted in to it.
- Arrangement of various components made simple to utilize every possible space.
- Ease of maintenance and servicing achieved by means of simplified layout that enables quick decision assembly of components.
- Scope of future improvement.

4.1 Mechanical Design

In mechanical design the components are listed down and stored on the basis of their procurement in two categories,

- Design parts
- Parts to be purchased.

For designed parts detailed design is done and dimensions there obtained are compared to next dimensions which are already available in market. This simplifies the assembly as well as the post production and maintenance work. The various tolerances on work are specified .The process charts are prepared and passed to manufacturing stage. The parts to be purchased directly are selected from various catalogues and are specified so as to have case of procurement.

Following are the various design components:

- Design of input Shaft.
- Design of output Shaft.
- Design for Selection of Ball Bearing
- Design of Load Drum Hub
- Design of Screw
- Brake Cable Failure Analysis

5. Summary

Testing of the E-barke system for trolley and cable to determine

1. Self-locking by proper selection and different pitch angles.
2. Mathematical model of dual worm system for optimal load lifting capacity, optimal factor of safety.
3. Derivation and resolution of system forces by drawing free body diagram of linkage.
4. Determination of forces and utilizing system of forces to determine the linkage dimensions of critical parts of drive.

References

[1] Surg min moon, Jaemging hue, Dachie Horg, Samghoon Lee, Chang sooltars, "Vertical motion contol of building façade maintance contact with built in guide rail ",(Robotics and computer integrated manufacturing 31 (2014).

[2] Boren at.al," United state patend ",(August, 7 2012)

[3] Otterson et al.," United state patend ",(August, 7 2012)

[4] Andrew Diehl, Product development director of Holmer solution, " Loads in zip STOP braking system",(Zip STOP released may 2013).

[5] Sivarao m. amaranth M.S. rizal and a camely ,"An investigation toward development of economical brake lining wear alert system",(International Journal of Engineering and Technology IJET-IJEN5, vol.09 Oct.2009)

[6] Tretsiak, Dzmitry, Kliauzovich, Siarhei, "Reasearch in self boosting disc brakes for commertial vehicles",(Belarusian national technical university, Belarus F2006SC04)

[7] United state patend, " Trolly Braking System", (U.S. application ser.no.12/411,253, filed mar.25, 2009)

[8] Erik Hultman, Mats Leigon, "A Cable feeder tool for robotized cable winding", (Robotics and computer integrated manufacturing 30 (2014)

[9] Young ok Lee, Minseok Jang, Wongoo Lee Choong Woo Lee, Chung Cho Chung, Young Seop Son, "Novel clamping force control for electric parking brake system" ,(Mechatronics 21 (2011) 1156-1162)

[10] Christopher Leonard James Line, "Control of an electro mechanical brake system" (2010)

19

REVIEW ON OVERLOAD TORQUE LIMITER WITH ELECTROMECHANICAL CLUTCH FOR TIMER BELT SPINDLE DRIVE

Chaitanya G Burande[1], Sanjaysingh Vijaysingh Patil[2]

[1]*PG Student, burandechaitanya5@gmail.com*

[2]*Assistant Professor, sanjaysingh.patil@sinhgad.edu*

Author Correspondence: Sai Shilpa society, Bibvewadi, Pune.

Abstract

Positive clutches used to transmit power between coincident shafts. The positive Engagement between clutch element ensure 100% torque transmission but occasionally the output shaft may be subjected to sudden overload which May make driving motor or engine to stall, which will lead to burn out electric Motor. In extreme cases this overload will lead to breakage of drive elements or clutch itself. In order to avoid this damage, it is needed that the input and output shafts be disconnected in case of sudden overload.

Torque limiter is overload safety devices which provide the reliable overload protection. Torque limiter is tamper proof. Generally whenever overload occurs in any shaft drive mechanism there is failure of following components possible-

1) Shaft / Coupling / Belt Drive
2) Machine shaft
3) Motor (Burning of electrical motor) Due to Overloading

In any case this damage leads to Replacements of machine parts and Ultimately
Replacement cost increases. To overcome all this problem available in machine element, Torque limiter will help us to avoid this damage.

Keywords: Torque limiter, eletromechanical clutch, timer belt spindle drive.

1. Introduction

Safety ball clutch are Overload Safety Devices with Torque Limiters which provide reliable overload protection. When a jam-up or excessive loading occurs the Safety ball clutch will reliably and quickly release to prevent system damage.

- These torque limiters are tamper-proof. Once installed, the torque value cannot be changed. This is an important feature that ensures the integrity of the machine design. Costly and potentially risky calibration procedures are not necessary. The torque value is controlled by the part number that is ordered. That value determines what spring is used during the assembly at the factory.
- The torque value can be changed in the field, however; the Safety ball clutch must be disassembled and the springs replaced to achieve the new torque value.
- Standard Safety ball clutch are bidirectional. The torque value is the same regardless of rotation. If specified, these torque limiters can be configured at the factory to release at different torque ratings for different rotational directions.

- In the coupling configuration, the Safety ball clutch fulfills two functions: (1) A flexible shaft coupling (2) a mechanical torque limiter. The Safety ball clutch in the shaft to shaft configuration will handle angular shaft misalignment up to 1.5 degrees and a .005" to .015" maximum parallel misalignment.
- The enclosed design of the mechanical Safety ball clutch enables it to operate in a wide variety of industrial environments. Special designs and materials can be made to withstand even more adverse conditions.

The torque value is determined by the force of the springs that are installed in the unit. The spring force acts upon the slides that are part of the inner shaft. These slides transmit force that will hold the drive key into an engagement slot in the outer housing. When the torque load exceeds the rating, (determined by precision tempered torque springs) the Safety ball clutch's drive key will pivot out of the engagement slot to disengage the Safety ball clutch. After disengagement the Safety ball clutch does not have significant resistance to rotation. Upon completion of one shaft rotation the Safety ball clutch will automatically try to reengage. Once the overload is removed and speed reduced, the drive key will snap into the engagement slot and the Safety ball clutch will be reset for the next overload event .This particular design is suitable for a single set of the design torque, and is not suitable for applications where the set torque may vary as per application. Rarely occurring overloads must be considered during the design process of a power train. These overloads can be evoked by malfunctions in the electronics of inverter and installation control, by obstructions in the work flow, by mis-operation etc. These overloads may create pre-damages which lead to full failure of the assembly or its components. Such affecting loads are avoidable by means of overload clutches. Thus reliable overload clutches are of strongly increasing interest for years.

1.1 Disadvantages of current system of overload protection

To protect the drive from failure What is available in market is a 'Flying ball clutch' which transmits torque from input to output using balls held by a spring in assembly when overload occurs the balls will come out of assembly –thus disconnecting input and output thereby saving part failure But

a) Rating of clutch is 1N-m, 5 N-m, 20 N.m etc. i.e fixed value so if o/p torque change we have to replace clutch.
b) Every time ball comes out of assembly we have to remove the clutch to replace ball this increases down time of machine
c) Drive always remains coupled there is no flexible arrangement like automobile clutch i.e. possibility to disengage at will.
d) If temporary overload occurs the clutch will slip and remain disengaged till it is preset even though the overload is now removed this leads to process down time.
e) Thus there is a need of Timer belt spindle drive with overload Safety ball clutch with following features

1.2 Advantages of overload torque limiter

- Electromechanical dis-engagement so that drive can be temporarily disengaged for I in process inspection or other activity.

- The Safety ball clutch can be set over a range of torques (say 0 to 20 kg-cm) so that the machine operator can set it to desired value for given application unlike the conventional clutches that are factory set.
- The transmission elements ie, the balls will not come out of assembly when there is overload slipping this comes as an advantage as the clutch can be preset without removing it from assembly this will save considerable amount of downtime of process as compared to the conventional clutch.
- If temporary overload occurs the clutch will slip and remain disengaged only till the overload is removed thus if the overload is removed while in running condition the clutch shall automatically engage and start transmitting power.

1.3 Problem Statement

Whenever an overload occurs in any shaft drive mechanism as shown below there are three possibilities:

a) Shaft / coupling/ belt drive fill fail or break.

b) Application i.e. machine shaft will fail or break

c) Motor will be overloaded resulting into electric burn…

In any case it is damage leading to machine part replacement …Down time of machine and increased part replacement and maintenance cost …

Machine tool spindle drives use timer belts with timer belt pulleys in open belt drive system. These spindle drives are as shown Fig 1.

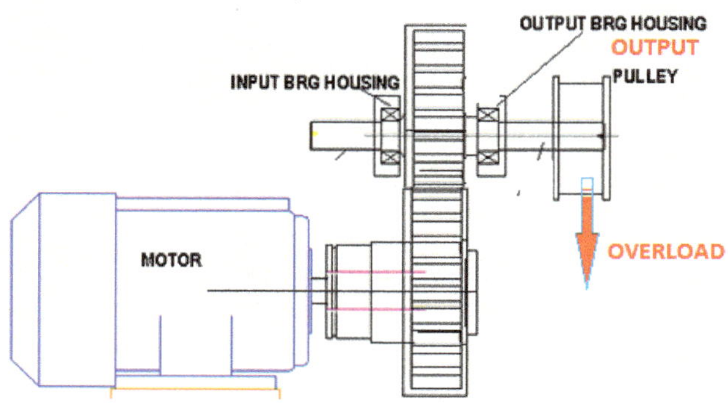

Figure 1: Machine tool spindle drives use timer belts with timer belt

2. Literature Review

"Transilvania University Brasov, Romania" in the paper titled "Design Procedure of Elastic and Safety Clutches using Cam Mechanisms" states that, topological and structural generation of elastic and safety coupling, In the second part, on the basis of the functional characteristic of the cam gears, we propose a simple method for the structural and generation of elastic and safety couplings[1].

"Nicolae EFTIMIE" in the paper titled "Dynamic Simulation Of The Safety Clutches With Balls" states that, explored the clutches are used largely in machine buildings, and by the correct selection of these depends – to a great extent – the safe and long working, both of these and of the kinematic chain equipped with them[2].

"M Jackell, J Kloepfer, M Matthias and B Seipel" in the paper titled "The novel MRF-ball-clutch design – a MRF-safety-clutch for high torque applications" states that the development of a safety-clutch by using magneto rheological fluids(MRF) to switch the transmission torque between a motor and a generator in a bus-like vehicle[3].

"Mr..S.Jegadeesan, Asst.Professor, N.Suganthi, M.E (Applied Electronics) V.S.B. Engineering College, karur" in the paper titled "Design of Energy Savings in Metropolitan Railway Substations and Communication Based Train Control" states that, explored the reduction in energy consumption has become a global concern and the EU is committed to reducing its overall emissions to at least 20%[4].

3. Design and Development

- System design as to number of ball-springs for desired torque capacity.
- Design and geometrical derivations of the groove profile in input base flange.
- Design and geometrical derivations of spring plunger profile.
- Selection and geometrical profile of clutch body ball holder.
- Selection and design of torque control using plunger and casing arrangement.
- Selection of solenoid coil for transmission of desired power
- Selection of timer belt drive for open belt drive
- Mechanical design : This includes the design and development of springs selection of suitable drive motor , strength analysis of various components under the given system of forces
- The critical components of assembly input pulley, solenoid mount, Safety ball clutch input shaft, input base flange, plunger, cylindrical body, output shaft etc., components will be designed using conventional theories of failure using various formulae, 3-D models of the above parts will be developed using Unigraphix software and meshing –analysis will be done, the result of stress produced will be validated using ANSYS-Workbench 14.5 release.

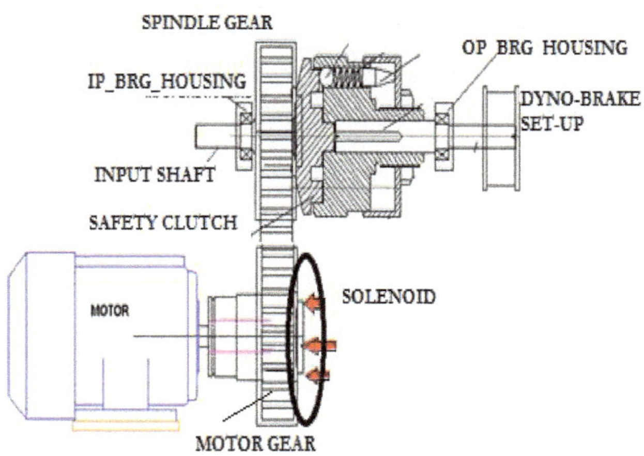

Figure 2: Experimental Setup

4. Summary

The torque value is determined by the force of the springs that are installed in the unit. The spring force acts upon the slides that are part of the inner shaft. These slides transmit force that will hold the drive key into an engagement slot in the outer housing. When the torque load exceeds the rating, (determined by precision tempered torque springs) the Safety ball clutch's drive key will pivot out of the engagement slot to disengage the Safety ball clutch. After disengagement the Safety ball clutch does not have significant resistance to rotation. Upon completion of one shaft rotation the safety ball clutch will automatically try to reengage.

Once the overload is removed and speed reduced, the drive key will snap into the engagement slot and the Safety ball clutch will be reset for the next overload event .This particular design is suitable for a single set of the design torque, and is not suitable for applications where the set torque may vary as per application .Rarely occurring overloads must be considered during the design process of a power train. These overloads can be evoked by malfunctions in the electronics of inverter and installation control, by obstructions in the work flow, by mis-operation etc. These overloads may create pre-damages which lead to full failure of the assembly or its components. Such affecting loads are avoidable by means of overload clutches

References

[1] I. Stroe, "Design Procedure of Elastic and Safety Clutches using Cam Mechanisms", 12th IFToMM World Congress, Besançon (France), June18-21, 2007.

[2] Nicolae Eftimie, "Dynamic Simulation Of The Safety Clutches With Balls", facility of technical science, NOVI SAD, May 18 2007.

[3] M Jackel 1, J Kloepfer, M Matthias and B Seipel, "The novel MRF-ball-clutch design – a MRF-safety-clutch for high torque applications", Journal of Physics: Conference Series 412 (2013) 012051.

[4] Mr.S.Jegadeesan, Asst.Professor, N.Suganthi, "Design of Energy Savings in Metropolitan Railway Substations and Communication Based Train Control", International Journal of Advanced Scientific and Technical Research Issue 3 volume 1, January-February 2013.

[5] STROE Ioan , "Elastic And Safety Clutches With Intermediate Rubber Elements", Fascicle of Management and Technological Engineering, Volume IX (XIX), 2010, NR1.

[6] Niranjan Singh, "General Review Of Mechanical Springs Used In Automobiles Suspension System", International Journal of Advanced Engineering Research and Studies.

[7] Drăghici, I,s.a., "The calculus and the building of the clutches", Editura Tehnică, Bucure ti, 1978.

20

Review Paper on Reliability Improvement Warranties

Madan M. Jagtap[1], Ashir Hawaldar[2], Niraj Jgtap[3], Saurabh Jain[4]

[1]*Saraswati College of Engineering, India, jagtap.aero@gmail.com*

[2] *Sarasvati College of Engineering, India, ashirhawaldar@gmail.com*

[3] *Sarasvati College of Engineering, India, niraj250819@gmail.com*

[4] *Sarasvati College of Engineering, India, Saurabh.jn1991@gmail.com*

Abstract

An automobile with thousands of parts and interactions among them is a highly complex product. It makes detailed testing/analysis during product development, manufacturing, and assembly a prohibitively enormous, if not infeasible, task. Thus, when the vehicle is put on the market, the likelihood of unexpected poor quality and reliability resulting in high warranty costs is not uncommon. Success during product development, manufacturing, and assembly is often judged by lack of quality and reliability

Key words: Warranty, Reliability, RIW.

1. Introduction

The reliability improvement warranty is a class of warranty

Policies those are applicable for expensive products that are especially built for customers. The products can be divided into two groups: (i) new acquisitions by defense involving new state-of-the-art technologies, and (ii) complex systems (such as locomotives, power generating units, etc.) which are custom built to serve the needs of commercial or industrial customers. This requires assessing reliability at different stages of the process and modifications to improve reliability if the agreed performance measures are not achieved. The measure usually used to assess product reliability and improvements there in is the mean time between failures (MTBF), and most RIWs include an MTBF. However, RIW can include any agreed-upon measure of reliability. As a result, RIW policies are often quite complex. In contrast to a consumer-type warranty, where the manufacturer decides on the warranty policy based on several factors, as each RIW is a unique warranty contract, often covering only a few items and carefully negotiated by the customer and the manufacturer. As a result, the cost analysis (for both manufacturer and customer) of each RIW policy poses new challenges, depending on the reliability terms included in the policy. In this paper, we deal with the major issues associated with RIW policies. The outline of the paper is as follows. we take a more detailed look at some of the specific features of such policies. Where we define the six different stages of the process and compare it with that for products sold with non-RIW policies. Finally, it is worth noting that in the RIW literature, the manufacturer is often referred to as "contractor" and the customer as "buyer" as the buyer contracts with the contractor to build the product. MTBF guarantee, logistics and other cost guarantees, mean time to repair, and similar guarantees Special features agreements – covering guarantees on characteristics such as ultimate life, Commercial service life, software, test and repair improvement, and other unusual or unique features.

RIW CONCEPT

RIW is a broader term that includes any warranty or guarantee that includes to improve product performance, reliability is product performance agreement (PPA). This class of warranties includes many types of contracts, RIW being just one warranty in one of four classes: Federal acquisition regulation agreements – covering inspection, supplies, design, performance specification, and technical data Contractor repair agreements – covering re-warranty of repaired or overhauled equipment, reliability guarantees, repair/exchange agreements, and RIW Field measurements agreements – covering MTBF verification tests, availability guarantees, RIW with MTBF guarantee, logistics and other cost guarantees, mean time to repair, and similar guarantees Special features agreements – covering guarantees on characteristics such as ultimate life, commercial service life, software, test and repair improvement, and other unusual or unique features.

RIW FEATURES

The original intent was to apply RIW in situations where operational (field), reliability, support costs, and potential reliability growth were all reasonably predictable. In such applications, many versions of RIW, with a wide variety of terms and features, were developed. Characteristics common to almost all applications include the following:

Complex equipment Clearly defined performance requirements Ability to evaluate field or operational performance Potential for reliability growth Contractor provision of field maintenance and repair Relatively long warranty terms (typically three to five years or more) Requirement for engineering analysis of failures Requirement for design changes to correct defects A fixed price contract Contractor fees based on demonstrated reliability improvements. Each bidding situation involving RIW may include many additional unique characteristics. Reliability, maintainability and supportability performance terms in an RIW policy may include one or more of the following items:

A guaranteed mean time between failures (MBTF)

A guaranteed turnaround time (TAT) for repaired or replaced units A supply of consignment spares for use by the buyer at no cost until the guaranteed MTBF is demonstrated Accuracy of testability System mission availability (point or interval availability).

ASSURANCE VERSUS INCENTIVE WARRANTIES

In case of assurance warranty performance meets some minimum level. In this the aim of the reliability improvement is to ensure that this is achieved. In contrast, in some cases the buyer is interested in encouraging the contractor to exceed the minimum level and as such the warranty contract includes incentive features to achieve this. This is accomplished by tying the payment to the performance level achieved by the contractor.

2. RIW PROCESS

The RIW process is different from that for items sold with non-RIW policies. There the manufacturer makes all the decisions regarding warranty terms. The buyer is directly involved only in that a selection is made among products with different warranties, although this could have an indirect effect in that adverse selection in sufficient numbers may influence the manufacturer to improve warranty terms. In contrast, in the case of products sold with RIW policies, the buyer is actively involved in the process from the beginning and, in fact, may influence some crucial warranty decisions, since warranty terms are usually negotiated by the two parties. In addition, the focus of RIW is on the total life cycle of the product rather than just the warranty period as in the case of non-RIW policies. As a result, the buyer has a strong input into product development and the nature of post-sale support needed subsequent to putting the product into operation. The RIW process involves six stages.

Stage 1 [Bid Proposal]: A bid is submitted in response to a Request for Proposal.

Stage 2 [Design and Development]: It is important to note that the contract might specify the type of testing to be carried out and the performance requirements to be met before the contractor can proceed to the next stage.

Stage 3 [Production]: As before, the contract might specify the type of testing to be carried out and the performance requirements to be met before the item is released for field operation.

Stage 4 [Field Operation]: Here the items are put into use. The buyer is required to ensure that the items are used in a manner stated in the contract. Should the operating environment deviate from that for which the product has been designed (that can arise later for example, the on some or all of the components of an item exceeding the design envelope), loads then it needs to be documented by the buyer and reported to the contractor. Also, the buyer might need to record other information that will be of assistance in fixing Problems on.

Stage 5(a) [Repair of Failed Items]: This requires planning of various resources, including spares, workshop facilities, technicians, and so forth.

Stage 5(b) [Engineering Change Proposals (ECP)]: This involves identification of failures and their causes and of modifications needed to prevent their future occurrence.

Stage 6 [Implementation]: This is concerned with actions required by Change Orders

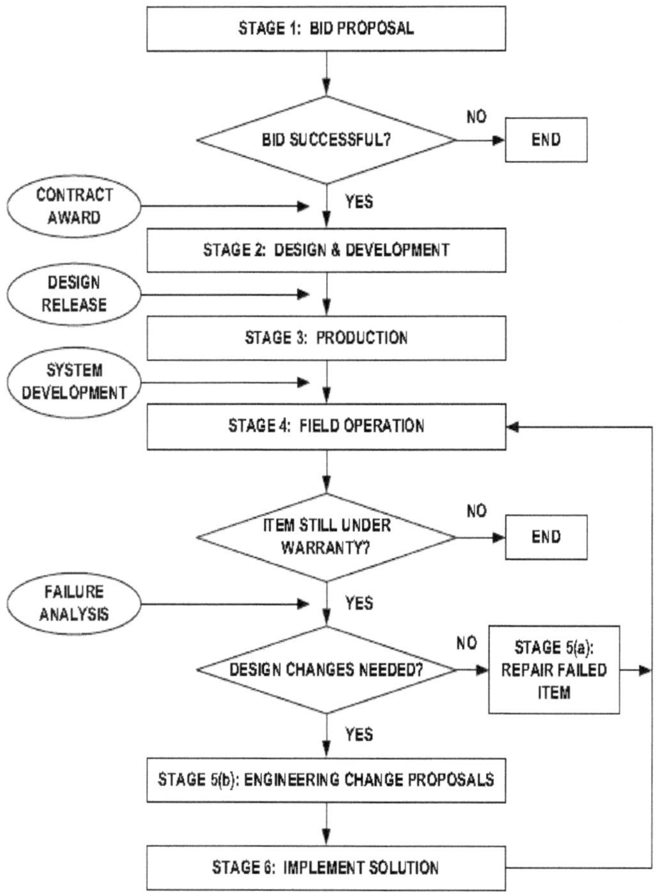

Fig Stage I Bid proposal

Bid Proposal [Stage 1]

The first stage of the RIW process is the bidding stage. A typical scenario begins with the buyer issuing a "Request for Proposal" (RFP) to one or more manufacturers, informing them of the specifications, including performance requirements, for the new product, along with other information needed to prepare a bid on the project. Manufacturers respond through an initial bid proposal that indicates how the new product can be realized and gives some indication of the performance levels and crude estimates of various costs. This is done using crude models based on limited data. The buyer carries out an evaluation of the proposals and decides on which ones are to be rejected and considers possible revisions to the performance requirements based on the bid proposals. The proposals not rejected go through a second stage where the process is repeated. The end outcome is the final contract that states the terms of the RIW policy.

1 CONTRACT

The contract needs to address the following issues.

What is the duration of the warranty period?

Is it an assurance or incentive warranty?

What are buyer and contractor obligations?

What issues are covered?

What are the exclusions?

How are the spare parts issues addressed? Are they delivered at the start

a) WARRANTY STATEMENT

This indicates which of the characteristics discussed in included. One or more of the following questions are addressed:

What is the duration of the warranty period?

Is it an assurance or incentive warranty?

What are buyer and contractor obligations?

What issues are covered?

What are the exclusions?

How are the spare parts issues addressed? Are they delivered at the start or

Over the operating life cycle of the product?

b) PRODUCT PERFORMANCE

The performance of the product must be stated properly so that there is no scope for ambiguity. Reliability-related performance needs to include the time frame for data collection, the type of data to be collected, and the procedures to assess performance in terms of the data. In the case of MTBF, one must specify whether a point or interval estimate is to be used. Similarly, during development the contract needs to indicate the kind of testing to be carried out and how to translate the test data into assessing performance at component, sub-system or system level. If these are not done properly, it can lead to disputes and litigation at a later time.

2 COSTS

There are several different types of costs involved. These include the following:

Development cost

Production cost

Support cost (for spares, etc.)

Warranty cost.

The contractor needs to take into account all of these costs in the pricing of the contract. The warranty cost must include the provision of replacements, repairs as well as upgrades. This depends on the duration and other terms of the warranty. Models are needed to predict these costs. From the buyer's perspective, the cost of interest is the total life cycle cost. The buyer might look at alternate warranty options and needs to assess the benefits of warranty traded against the warranty costs. All activities needed to build and support the product and, all the major factors that influence these activities must be converted into costs that are projected far enough into the future to cover the warranty period

3 RISKS

The contractor faces several kinds of risk. These include the following:

Technical risk: This results from not achieving the performance levels stated in the contract and, as a consequence, incurring a large penalty through high warranty costs and/or high cost of engineering design modifications

Project risk: This results from not delivering the product in time and/or cost overruns during development and production.

4 DISPUTE RESOLUTIONS

For most RIW policies, the product as well as the contract is complex. This implies that the contract might not address some issues that can lead to potential problems and disputes after the contract has been signed. Also, the interpretation of the contract (for example, the testing conditions or operating environment) and other unverifiable factors (for example, the cause of failure being either due to operator error or design weakness) can lead to possible conflicts. As such, both parties (buyer and contractor) need to look at alternate dispute resolution mechanism during Stage 1.

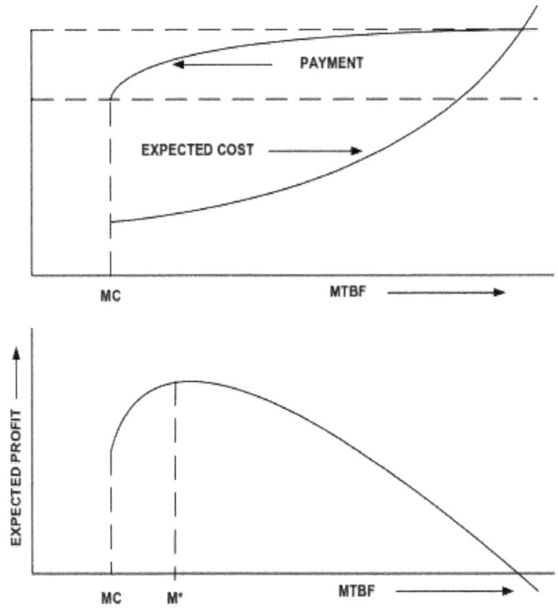

Fig Optimal MTBF under incentive RIW

MODELS

The RIW process is complex and varies considerably from contract to contract. As a result, quite sophisticated models are needed to capture all of the elements of RIW. The models need to include development, testing, and operational factors and related costs, reliability growth, randomness (uncertainty of outcomes), risk, other project-specific elements, and, in addition, must look at these from a life cycle perspective and from the points of view of both buyer and contractor. General models that embody all of these features have not been developed. Typically, what is done is to develop comprehensive computerized databanks to record as much information as possible about the RIW project.

EXAMPLES

Most of the examples are from the defense sector. A few examples of applications7 for which models were developed are:

PURCHASE OF AVIONICS BY THE MILITARY.

A batch of items was purchased, with a subset of these put into service and the remainder used as spares. The model expressed cost to the buyer as a function of purchase price, expected cost of failures, and cost of maintenance, and included a number of breakdowns into lower level cost elements, incentives for reliability improvement, and other relevant factors.

MILITARY ELECTRONIC EQUIPMENT: The model considered cost of RIW with guaranteed spares availability and turnaround time, and involved many related factors.

ECP AND IMPLEMENTATION

Although the predicted performance during design and development and during production tests may meet or exceed the stated performance measures, this does not guarantee that the desired performances are achieved during field operation. This is because of the limitations of models used prior to and during testing and of the data available from tests. However, if the root cause analysis indicates a major potential problem or that the performance terms are not being met, then the contractor needs to provide the buyer with engineering change proposals that will overcome the problem and ensure that the performance levels are met. This involves a proper evaluation of the modified design and in some cases test data to prove the claims of the contractor. Once the buyer gives approval, the contractor needs to carry out changes on all the items delivered or being held as spares.

It is worth noting that if an item is used in a mode that the product has not been designed for (for example, running a locomotive at speeds higher than designed or on poorly maintained tracks), then the warranty would become null and void. This implies that the buyer needs to ensure that proper operational and maintenance related data are kept in order to verify, if necessary, that the buyer's obligations are met.

3. MANAGEMENT OF THE RIW PROCESS

Some of the management issues for RIW are the same as for items sold with non-RIW policies and discussed in Chapter 4. However, there are some issues that are unique to RIW policies and we discuss these briefly in this section.

1 WARRANTY NEGOTIATIONS

This is a critical element of Stage 1 for both commercial and government acquisitions. The negotiation process serves to clarify issues so as to arrive at an agreement that is acceptable to both parties (contractor and buyer). The negotiations can be involved, as the interests of the two parties are different and so are the information and knowledge needed to evaluate costs and risks.

2 PROJECT MANAGEMENT

For a complex product, the contractor needs the involvement of several subcontractors to provide different sub-systems. For example, in the case of an aircraft, the engine manufacturer would be one of the main sub-contractors. It is the responsibility of the contractor to confirm that the components obtained from subcontractors conform to specifications. This implies that all design changes must be communicated effectively.

3 DATA MANAGEMENT

The contractor needs a system to track and collect relevant data needed to estimate achieved performance at different stages of development, during production and during field operation. The input to this database must include the following: Product design: detailed design specifications of original configuration; design changes based on field performance data Product operation: usage, failures, failure modes, failure times, etc. Repair: actions, time needed for repair, costs, etc. The buyer needs to keep track of operational-related data to ensure that the terms of the contract are not violated.

4 WARRANTY ADMINISTRATIONS

This deals with systems and procedures required to execute the various stages indicated in government acquisitions. It is important to specify requirements, procedures, and criteria for assessment of performance and to perform a careful cost analysis at the outset. In the bid preparation stage, careful attention must be paid to all of these factors as well as to the cost and feasibility of various warranty terms. The last is especially important since a key feature of RIW acquisitions is that warranty terms will be negotiated. Brennan offers many guidelines regarding bid preparation, including discussion of Risks and benefits Warranty trade-offs

Warranty costing

Performance guarantees

Warranty negotiating

Determination of cost effectiveness.

4. CONCLUSION

RIW is can be used very effectively to reduce warranty cost related with different products

REFERENCES

[1] Blischke, W. R. and Murthy, D. N. P. (1994), Warranty Cost Analysis, Marcel Dekker, New York.

[2] Wu S, Li H. Warranty cost analysis for products with a dormant state. European Journal of Operational Research 2007; 182: 1285-1293.

[3] Hiller, G. E. (1973), "Warranty and product support. The plan and use thereof in a commercial U.S· Navy Aviation Supply Office, Philadelphia.

[4] Wu S, Xie M. Warranty cost analysis for no repairable services products. International Journal of Systems Science 2008; 39: 279-288.

[5] Murthy DNP, Solem O, Roren T. Product warranty logistics: Issues and challenges. European Journal of Operational Research 2004; 156: 110-126.

[6] Karim MR, Suzuki K. Analysis of warranty claim data: A literature review. International Journal of Quality and Reliability Management 2005; 22: 667-686.

[7] Murthy DNP. Product warranty and reliability. Annals of Operations Research 2006; 143: 133-146.

Supplier-Oriented Knowledge Management in the Automobile Industry

Hrushikesh D. Sawant[1], S.N.Teli[2], L.M. Gaikwad[3]

[1]*PG-Student, Mech. Engg. Dept. SCOE, Kharghar, Navi Mumbai, Maharashtra*

[2] *Associate Professor & Head, Mech. Engg. Dept. SCOE, Kharghar, Navi Mumbai, Maharashtra*

[3]*Asst. Professor, Mech. Engg. Dept. BVCOE, Belpada, Navi Mumbai, Maharashtra*

[1]*hrushielectra@gmail.com*, [2]*shivanandteli@yahoo.com*, [3]*lokpriya2004@yahoo.co.in*

Abstract:

This study focuses on the impact of knowledge management in supplier relations within the automobile industry, on financial and product performance. A model of the relationship between supplier-oriented knowledge management, including its associated processes, and performance is described. Knowledge sharing through face-to-face communication is positively related to both product and financial performance, while technological knowledge sharing has a positive impact on product performance under conditions of high technological dynamism. Supplier involvement in the production process is related to product performance and use of knowledge management tools is related to financial performance.

***Key words**:* Knowledge management, Supplier, Automobile industry

1. Introduction:

Knowledge, knowledge workers and Knowledge Management are topics receiving increasing attention from a variety of discipline. A recent survey by the Journal of Knowledge Management revealed that 92% of the responding executives indicated that they worked in knowledge intensive organizations. At the same time new technologies have been developed to better enable the management of knowledge. Knowledge Management is an emerging discipline as the potential uses, features and benefits of the current incarnation of Knowledge Management are still being defined and as increasing numbers of people and organizations begin to explore this new form of communication and organizational learning. Much work remains. From the Nonaka's model as shown in figure 1, KM is described as knowledge creation process, which represented by spiral conversion mechanism. We believe that each process (socialization, externalization, internalization, combination) managing organizational knowledge assets such as lessons learned, best practices, report, technical documents and so on. We can also elaborate the model by showing some examples of KM technologies that may be applied to facilitate the knowledge conversion process such as described in (Marwick 2001).

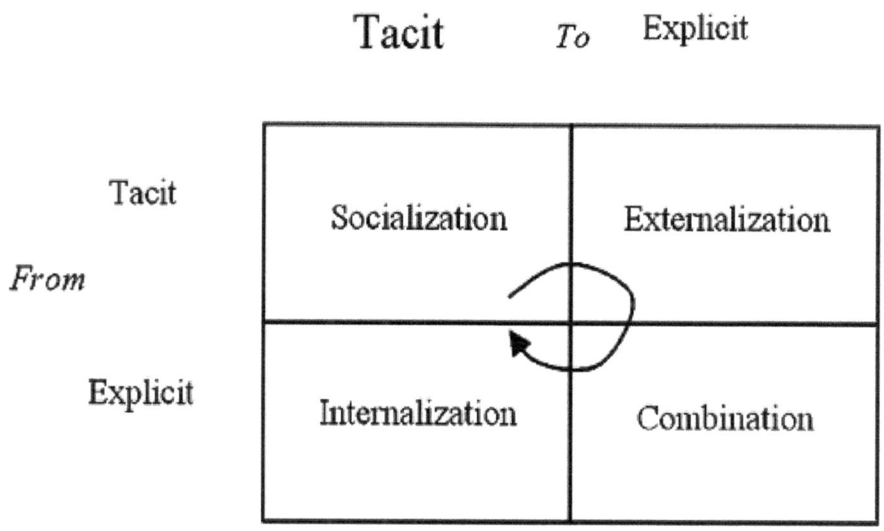

Fig. 1: Takeuchi's KM Model (1995)

While there are many processes of KM, such as gathering, searching, filtering, conceptualizing, projecting and transferring (Park and Kim, 2006), this study presents KM as a combination of five specific processes. These are:

1. Knowledge sharing and distribution
2. Knowledge generation and development
3. Knowledge codification and storage
4. Organization leader roles and
5. Reward systems.

2. Literature Review:

Daniel Palacios Marques et.al (2006) in their research work studied the connection between knowledge management practices and firm performance.

Kwong-Chi Lo, Kwai-Sang Chin (2009) This paper aims to develop a user-satisfaction-based knowledge management performance measurement model, including identifying the assessment criteria sourced from user-satisfaction-based core values, critical success factors and phases of knowledge management process. With the aid of the developed measurement model, organisations could assess the strength and weakness of their own knowledge management system and practice and then identify areas for improvement. User-satisfaction-based core values, critical success factors and five-phase knowledge management process are identified through literature review and analysis, and expert interviews, with the aid of system modelling techniques. The seven user-satisfaction-based core values, eight critical success factors and five-phase knowledge management process are identified as the basis of the assessment criteria.

M.D. Singh, Ravi Shankar et.al (2006) Knowledge management involves strategies and processes of identifying, capturing, and leveraging knowledge to enhance competitiveness. The new world of knowledge-based organizations is distinguished from the organizations of the last millennium by its emphasis on monitoring and controlling the organization by shared knowledge derived from internal and external data sources. It believes in continual transformation of the knowledgebase according to changing business strategy.

Balasubramanian et al. (1999) define knowledge management as an organizational capability that allows people in organizations, working as individuals, or in teams, projects, or other such communities of interest, to create, capture, share, and leverage their collective knowledge to improve competitiveness and performance.

Raisinghani (2000) conceptualizes the concept of knowledge management as the concern for the creation of structures, which combine the most advanced elements of technological resources and the indispensable input of human response and decision making, a view that resonates with the knowledge-based theory of the firm developed by Grant (1996).

Hansen et al. (1999) have identified two broad approaches to the management of knowledge in organizations, viz., the personalization approach and the codification approach. The personalization approach would include face-to-face communication, communication through such structures as networks of people, cross-functional teams, committees, task forces, training and development, internal knowledge sharing through benchmarking and job rotation, and creation of strategic alliances. The codification approach refers to the technological route for knowledge management and would include the setting up of databases, data warehouses, decision support systems, ERP systems, and electronic networks for communication and sharing knowledge.

3. Supplier oriented knowledge management:

In addition to the social and technological components of knowledge management within organizations described above, for the purpose of coordination, the knowledge management literature also indicates that organizations manage the process through which they acquire, and organize knowledge from outside the organization (from suppliers) and disseminate such knowledge and information within the organization (Alavi and Leidner, 2001; Teigland and Wasko, 2003). These and other researchers (Zack, 1999) suggest that the processes that are set up for the specific enhancement of the data and information obtained from various entities in the environment, such as suppliers, constitute a key component of knowledge management. Such processes, focused on suppliers, can be labelled supplier-focused knowledge management.

Although the literature on supplier relations in the auto industry has been developing for some time (Clark and Fujimoto, 1989; Dyer, 1996, 1997; Dyer and Nobeoka, 2000; Belzowski et al., 2003; Kaufman et al., 2000; Martin et al., 1995; Mudambi and Helper, 1998; Takeishi, 2001), the specific focus on collaboration for knowledge sharing purposes is fairly new (Belzowski et al., 2003; Dyer and Nobeoka, 2000; Takeishi, 2001). With the exception of the few studies that utilize this new focus, most of the others in the literature focus on supplier relations from perspectives such as transaction costs, collaborative versus adversarial relations, and variables such as asset specificity and location specificity. A focus on collaborative supplier relationships for knowledge sharing purposes has been argued to be a very important determinant of competitive advantage in the automobile industry (Dyer and Nobeoka, 2000), among others such as pharmaceuticals (Pisano, 1994), and hence the need to examine such knowledge management vis-à-vis suppliers has gained increased importance. We identify and focus on several processes involved in the management of knowledge vis-à-vis suppliers in the Brazilian automobile industry. We then empirically test for the impact of these processes, within the framework of our supplier-focused knowledge management model, on product and financial performance. Such empirical tests in the automotive industry exist only at a qualitative, case study level in this literature (Dyer and Nobeoka, 2000; Takeishi, 2001). We integrate this literature, provide a model, and thus make important contributions to this literature in the context of the Brazilian auto industry.

3.1 Supplier knowledge sharing:

Recent additions to the supplier relations literature within the auto industry have identified the importance of knowledge sharing and the building of a network for such purposes (Aoshima, 2002; Belzowski et al., 2003; Dyer and Nobeoka, 2000). Outside of these recent attempts, much of the rest of this literature does not directly examine knowledge sharing in supplier relations in the auto industry. However, this literature does provide indications to the effect that collaborative supplier relations and information sharing in a network mode are likely determinants of competitive advantage in this industry (Dyer, 1996, 1997; Mudambi and Helper, 1998). Dyer (1996) found evidence which suggests that specialized networks of suppliers provide a competitive advantage to manufacturers in the auto industry (Millington et al., 1998). In Dyer's (1996) study, one aspect of specificity, viz. Location specificity, was closely thought to be linked to the degree of information sharing between supplier and manufacturer. The degree of such information sharing is a crucial component of our concept of supplier-oriented knowledge management. Such information sharing is usually either assumed to exist (Dyer, 1996) in studies of the auto industry or qualitatively assessed in case study methodologies (Dyer and Nobeoka, 2000). We hypothesize for and directly measure such information sharing in this study as components\ of our concept of supplier-focused knowledge management, thus providing valuable additions to the literature.

In addition to information sharing, Dyer (1996) also emphasized the resulting benefit of conversion of tacit knowledge to explicit knowledge through these supplier networks, as an important element contributing to competitive advantage in the auto industry.

Dyer and Nobeoka (2000) focus on a rich qualitative investigation of the nature of the knowledge sharing practices at Toyota within their network and sub-networks. Such interpersonal processes of knowledge sharing (as opposed to technological routes) are more capable of converting tacit knowledge to explicit knowledge, a crucial component contributing to competitive advantage. Very little work exists along these lines of investigating multiple processes of knowledge sharing across organizational contexts in the auto industry (Aoshima, 2002). We include knowledge sharing, using both media rich, face-to-face communication, and through technological networks such as electronic data transfer and reliance on knowledge management tools, in our conceptualization and measurement of the knowledge sharing component of supplier-focused knowledge management. Thus, we conceptualize and operationalize both the personalization and codification approaches to knowledge management identified in the literature (Hansen et al., 1999). The central argument is that knowledge sharing is a crucial component of supplier-focused knowledge management and thus its impact on performance.

3.2 Supplier integration in the production process:

Although the supplier relations literature has mainly focused on the innovation benefits of close ties with suppliers in the product development process (Dyer and Singh, 1998; Kaufman et al., 2000; Pisano, 1994; Takeishi, 2001), such close relations with suppliers and integration into the production process also provides several benefits in the manufacturing realm (Millington et al., 1998). Evidence from the auto industry (Millington et al., 1998) suggests that the benefits of suppliers possessing knowledge of the manufacturer's production process and integration into their production provides benefits such as reduced costs in multiple areas, flexibility under varying demand conditions, reduced risk of disrupted deliveries, possibility of sequencing production of pre-assemblies and assemblies, and the potential for experiential learning by producing a focused set of components/products over the long run. Such benefits were identified by Millington et al. (1998) in the context of their investigation of 'local assembly units' near manufacturers. However, such local units achieved these benefits only by engaging in higher levels of interactions, resulting in higher knowledge of the production process and higher integration into the production process of the manufacturer.

These knowledge aspects were not assessed directly in this case study (Millington et al., 1998) but were assessed in (Dyer and Nobeoka's, 2000) case study of Toyota's network. The literature still lacks quantitative evidence of such supplier knowledge of the production process, and integration into the production process, and the resulting impact of these on product and financial performance. Increased knowledge of the interdependencies of the production process and of the interfaces in such processes spans across both manufacturing and product development processes, since these are fundamentally interlinked (Clark and

Fujimoto, 1989). Pisano (1994) provided evidence from the pharmaceutical industry for the positive link between knowledge and the product development process, especially in terms of the time for development. Within the context of the supplier-focused knowledge management model, increased levels of supplier knowledge of the production process is likely to help in their collaborative efforts with manufacturers. As suggested above, higher levels of supplier knowledge integration into the production process is likely to make these suppliers manufacturer chains more highly capable of structuring and organizing the production process to provide more flexibility and quickness and the competitiveness resulting from it (Takeishi, 2001). Higher levels of supplier involvement in the production process are also necessary to prepare them for higher levels of sharing knowledge and thus easier conversion of tacit knowledge about interfaces and interdependencies to explicit knowledge. Such conversion is crucial for firms trying to build specialized networks with suppliers to build their competitive advantage.

Information technologies and flexible manufacturing technologies have allowed for the inexpensive modification of production set-ups and sequences and thereby increase the inter-firm know-how of their specific interfaces (Dyer and Singh, 1998; Kaufman et al., 2000). Although these technologies increase the resource availabilities in these areas, the actual competitive advantages accrue only to those firms that have the capabilities to utilize these technology and knowledge resources, as evidenced by the motivational challenges, among others, faced by Toyota in building its supplier network to share knowledge (Dyer and Nobeoka, 2000) and the difficulties faced by North American firms to replicate such links (see Martin et al., 1995). Thus, it becomes imperative to investigate actual levels of supplier knowledge of the production process and their level of integration into the manufacturer's production process, because these are the crucial underlying variables that impact product and financial performance.

3.3 Supplier early involvement in product design:

In addition to supplier involvement in the production process, involvement in the earlier stages, viz. design, can improve crucial capabilities such as anticipating for and designing for tacit knowledge of interdependencies and for managing interdependencies in an optimal manner to enhance ongoing quickness of the process and its flexibility (O'Grady, 1999). The supplier relations literature has touched on the innovation benefits of close ties with suppliers in the product development process (Dyer and Singh, 1998; Eisenhardt and Tabrizi, 1995; Kaufman et al., 2000; Pisano, 1994; Takeishi, 2001). It has become common practice in the auto industry for suppliers to set up shop within the manufacturer's premises and to be involved in the project right from the design phase (Millington et al., 1998; Takeishi, 2001). Such practices maximize location specificity, which Dyer (1996) argued to be strongly connected to information and knowledge sharing. Such knowledge sharing is in the specific form of technical exchange and cooperation in the technological realm. Thus, early involvement of the suppliers in the product development and manufacturing process, in the form of collaborative design and in the form of technical exchange and cooperation, is a critical component of the supplier-oriented knowledge management.

The direct examination of early supplier involvement in design in terms of its impact on performance in the auto industry is rare (Takeishi, 2001). We contribute to the literature by directly examining this empirical relationship between early supplier involvement in design and product and financial performance.

The information systems literature on knowledge management provides an in-depth analysis of the impact of technological knowledge management tools on organization performance. Specific knowledge management systems, built on custom-designed knowledge management architectures (Zack, 1999) have been known to relate strongly to organizational performance (Armstrong and Sambamurthy, 1999; Davenport et al., 1998). The codification approach to knowledge management (Hansen et al., 1999) is entirely consistent with this view of the impact of the technological route to knowledge management. Our study suggest that supplier early involvement in product design in the form of higher levels of interaction between suppliers and manufacturers using technological knowledge management tools, as well as the early involvement of suppliers through co-design, will result in higher levels of product and financial performance.

3.4 Moderating factor: technological dynamism

The supplier relations literature (Dyer and Nobeoka, 2000; Sanchez and Mahoney, 1996) suggests that increasing levels of knowledge sharing in manufacturer- supplier combinations has an important impact on competitive advantage of organizations by providing the capability to more effectively handle environmental change. Higher levels of collaborative relationships focused on knowledge sharing provides firms with the capabilities to reduce costs, bring products to market quicker, stabilize and improve quality, and reduce risk of disruptions, all of which cushion the effect of environmental change. Conversely, the rate of environmental change could affect the relationship between supplier-focused knowledge management and product and organizational performance as identified by some in this literature (e.g. Clark and Fujimoto, 1989; Dyer and Nobeoka, 2000; Eisenhardt and Tabrizi, 1995; Kaufman et al., 2000; Takeishi, 2001). Such environmental change, under certain conditions, makes supplier-focused knowledge management more important and therefore more strongly associated with performance, than under other conditions. Specifically, supplier-focused knowledge management could be more important and have a stronger impact on performance under conditions of high environmental change than under low environmental change conditions (see Hitt et al., 2000, for a related argument and empirical finding). Hitt and colleagues (2000) found results supporting their argument that there is a generally a greater willingness to share knowledge in emerging economy enterprises than those in developed economies. Brazil's auto parts industry has been identified as a 'hot' sector with high growth rates. This, in combination with the presence of highly competitive steel, glass, and rubber industries, and massive investments to the technology sector of Brazil makes for a highly dynamic economic and technological environment (Hoskisson et al., 2000).

Following suggestions in the supplier-relations literature (e.g. Clark and Fujimoto, 1989; Dyer and Nobeoka, 2000; Eisenhardt and Tabrizi, 1995; Kaufman et al., 2000; Takeishi, 2001) we focus on technological dynamism as one of the major factors that results in environmental uncertainty and creates the need for strategic flexibility. When competitors introduce new products using new technologies at a rapid pace, thereby increasing the level of product variety, it forces all manufacturers to respond strategically or put themselves in the strategic position of being able to respond to such changes. Knowledge management in general and supplier-focused knowledge management in particular serves to provide this capability of flexibility and quick response. Thus, it stands to reason that such knowledge management becomes more important under conditions of high technological dynamism than under conditions of low technological dynamism. Therefore, we included the level of technological dynamism measured as the rate of technological change as a relevant moderating variable that impacts the relationship between supplier-focused knowledge management and firm performance. Additionally, under conditions of rapid technological change, the conversion of tacit knowledge to explicit knowledge becomes more challenging and calls for increased levels of interaction, collaboration, cooperation, and knowledge sharing. This is because the bases of both tacit and explicit knowledge change with the rapid changes in the technology. Thus, rapid changes in product design and the manufacturing process, as is the case in the auto industry (e.g. Clark and Fujimoto, 1989), necessitates higher levels of knowledge management between suppliers and manufacturers, without which the manufacturer's strategic flexibility and responsive ability is likely to suffer. Eisenhardt and Tabrizi (1995) found some evidence of the moderating impact of uncertainty (in the form of unpredictable products) on the relationship between supplier involvement and product performance (in the form of speed of product development). The above suggests that technological dynamism as measured by the industry rate of technological change is likely to moderate the relationship between supplier-focused knowledge management and performance. This moderating impact is likely to apply to each of the components of supplier-focused knowledge management.

4. A Model of Supplier-Oriented Knowledge Management:

With the help of knowledge management and supplier-relations literatures the framework of supplier-oriented knowledge management is built. In addition to these literatures, we also draw from a recent qualitative study (Kotabe et al., 2007), in developing the model and its hypotheses. These literatures suggest that: (a) supplier knowledge sharing (e.g. Aoshima, 2002; Belzowski et al., 2003; Dyer and Nobeoka, 2000); (b) supplier knowledge integration to the production process (e.g., Millington et al., 1998); and (c) early supplier involvement in product design through the use of knowledge management tools (e.g. Dyer and Singh, 1998; Eisenhardt and Tabrizi, 1995; Kaufman et al., 2000; Pisano, 1994; Takeishi, 2001) are the three dimensions of supplier focused knowledge management. The three dimensions indicated here are suggested in the literature mentioned above but have not been directly examined in this literature. Moreover, most of this literature is qualitative and thus the relationships remain to be tested (directly or indirectly). We blend these three dimensions into our model of the relationship between supplier oriented knowledge management and performance, in the context of the industry and its customers and present it in Figure 2. The model in Figure 2 indicates a direct and positive relationship between supplier-oriented knowledge management and financial and product performance. We include technological dynamism as a moderator variable that affects the relationship between supplier-focused knowledge management and performance.

The central thesis of the supplier-oriented knowledge management model we develop here is that it has a favourable impact on product performance as identified by its various characteristics such as cost, speed to market, quality, and reputation, in addition to its independent effect on financial performance. All the three components in our model enhance collaboration between suppliers and manufacturers with the focus on sharing and managing knowledge.

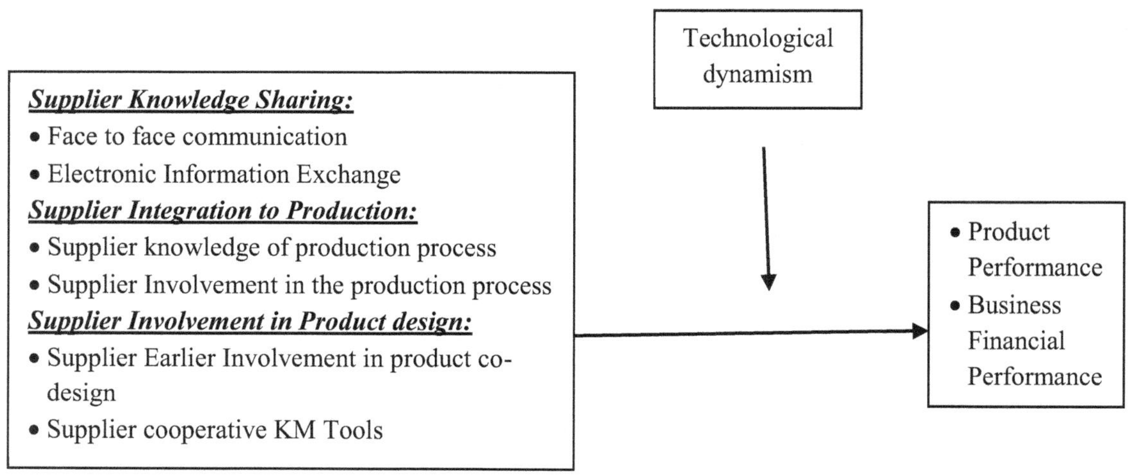

Figure

Figure 2 A model for supplier oriented Knowledge Management

5. Conclusion:

Following are the conclusions drawn from the study of supplier-oriented knowledge management system:

- The new world of knowledge-based organizations is distinguished from the organizations of the last millennium by its emphasis on monitoring and controlling the organization by shared knowledge derived from internal and external data sources.
- It believes in continual transformation of the knowledge-base according to changing business strategy.

- The objective of this paper is to understand the KM practices in an automobile industry in India, which are going through a major transition in this area. This paper presents knowledge management as a coordinating mechanism.
- Empirical evidence supports the view that a firm with a knowledge management capability will use resources more efficiently and so will be more innovative and perform better. The importance of training based on competency gap, SAP and R& D for enhancing the Knowledge of employee's is clearly pointed out in this research work.
- Knowledge sharing through face-to-face communication is positively related to both product and financial performance, while technological knowledge sharing has a positive impact on product performance under conditions of high technological dynamism.
- Supplier involvement in the production process is related to product performance, and use of knowledge management tools is related to financial performance.
- Supplier oriented knowledge management model that explicates the various means through which organizations in globally competitive industries can manage knowledge vis-à-vis external entities such as suppliers.

Specifically, our study has identified and presented evidence in favour of face-to-face knowledge and information sharing, electronic knowledge sharing, early involvement in product design, reliance on knowledge management tools for technological exchange, and supplier knowledge of manufacturer production processes.

References:

[1] Armstrong, C. P. and Sambamurthy, V. (1999). 'Information technology assimilation in firms: the influence of senior leadership and IT structures'. Information Systems Research, 10, 304–27.

[2] Alavi, M. and Leidner, D. E. (2001). 'Knowledge management and knowledge management systems: conceptual foundations and research issues'. MIS Quarterly, 25, 107–36.

[3] Clark, K. B. and Fujimoto, T. (1989). 'Lead time in automobile product development: explaining the Japanese advantage'. *Journal of Engineering and Technology Management*, **6**, 25–58.

[4] Daniel Palacios Marques, Fernando Jose Garrigos Simon(2006),"The Effect of Knowledge Management Practices on Firm Performance", *Journal of Knowledge Management,* **10**, Issue: 3, Page: 143 - 156.

[5] Davenport, T. H., De Long, D. W. and Beers, M. C. (1998). 'Successful knowledge management projects'. Sloan Management Review, 39, 43–57.

[6] Dyer, J. H. and Nobeoka, K. (2000). 'Creating and managing a high-performance knowledge-sharing network: the Toyota case'. Strategic Management Journal, 21, 345.

[7] Eisenhardt, K. and Tabrizi, B. (1995). 'Accelerating adaptive processes: product innovation in the global computer industry'. Administrative Science Quarterly, 40, 84–110.

[8] Hansen, M.T., Nohria, N. and Tierney, T. (1999). 'What's your strategy for managing knowledge?'. Harvard Business Review, 77, 106–118.

[9] Kwong-Chi Lo, Kwai-Sang Chin (2009), "User-satisfaction based Knowledge Management Performance Measurement", : International Journal of Quality & Reliability Management, 26, Issue: 5, Page: 449 - 468.

[10] Hitt, M., Dacin, M. T., Levitas, E., Arregle, J-L. and Borza, A. (2000). 'Partner selection in emerging and developed market contexts: resource-based and organizational learning perspectives'. Academy of Management Journal, 43, 449–67.

[11] Hoskisson, R. E., Eden, L., Lau, C. M. and Wright, M. (2000). 'Strategy in emerging economies'. Academy of Management Journal, 43, 249–67.

[12] Jenny Darroch (2005), "Knowledge Management, Innovation and Firm Performance", Journal of Knowledge Management, 9, Issue 3, pp.101-115.

[13] Kotabe, M., Parente, R. and Murray, J. (2007). 'Antecedents and outcomes of modular production in the Brazilian automobile industry: a grounded theory approach'. Journal of International Business Studies, 38, 84–106.

[14] M.D. Singh, Ravi Shankar, Rakesh Narain, Adish Kumar (2006), "Survey of Knowledge Management Practices in Indian Manufacturing Industries", Journal of Knowledge Management, 10, Issue: 6 Page: 110 – 128.

[15] Pisano, G. (1994). 'Knowledge, integration, and the locus of learning: an empirical analysis of process development'. Strategic Management Journal, 15, Special Issue, 85– 100.

[16] Sanjay Kumar Singh (2008), Role of Leadership in Knowledge Management: a Study", Journal of Knowledge Management, 12, Issue 4, pp.3-15.

[17] Zack, M. H. (1999). 'Managing codified knowledge'. Sloan Management Review, 40, 45–58.

Permissions

All chapters in this book were first published in IJRAME, by IJRAME; hereby published with permission under the Creative Commons Attribution License or equivalent. Every chapter published in this book has been scrutinized by our experts. Their significance has been extensively debated. The topics covered herein carry significant findings which will fuel the growth of the discipline. They may even be implemented as practical applications or may be referred to as a beginning point for another development.

The contributors of this book come from diverse backgrounds, making this book a truly international effort. This book will bring forth new frontiers with its revolutionizing research information and detailed analysis of the nascent developments around the world.

We would like to thank all the contributing authors for lending their expertise to make the book truly unique. They have played a crucial role in the development of this book. Without their invaluable contributions this book wouldn't have been possible. They have made vital efforts to compile up to date information on the varied aspects of this subject to make this book a valuable addition to the collection of many professionals and students.

This book was conceptualized with the vision of imparting up-to-date information and advanced data in this field. To ensure the same, a matchless editorial board was set up. Every individual on the board went through rigorous rounds of assessment to prove their worth. After which they invested a large part of their time researching and compiling the most relevant data for our readers.

The editorial board has been involved in producing this book since its inception. They have spent rigorous hours researching and exploring the diverse topics which have resulted in the successful publishing of this book. They have passed on their knowledge of decades through this book. To expedite this challenging task, the publisher supported the team at every step. A small team of assistant editors was also appointed to further simplify the editing procedure and attain best results for the readers.

Apart from the editorial board, the designing team has also invested a significant amount of their time in understanding the subject and creating the most relevant covers. They scrutinized every image to scout for the most suitable representation of the subject and create an appropriate cover for the book.

The publishing team has been an ardent support to the editorial, designing and production team. Their endless efforts to recruit the best for this project, has resulted in the accomplishment of this book. They are a veteran in the field of academics and their pool of knowledge is as vast as their experience in printing. Their expertise and guidance has proved useful at every step. Their uncompromising quality standards have made this book an exceptional effort. Their encouragement from time to time has been an inspiration for everyone.

The publisher and the editorial board hope that this book will prove to be a valuable piece of knowledge for researchers, students, practitioners and scholars across the globe.

List of Contributors

Supriya Burgul
PG Scholar (Mechanical), Vishwakarma Institute of Information Technology

Vishal N. Kshirsagar
P.G. Student, Department of Mechanical Engg, KDK College of Engineering, Nagpur, R.T.M. Nagpur University, Maharashtra, India

Dr. S. K. Choudhary
Asst. Professor, Department of Mechanical Engg KDK College of Engg. Nagpur, RTM Nagpur University, Maharashtra, India

Prof. A. P. Ninawe
Asst. Professor, Department of Mechanical Engg KDK College of Engg. Nagpur, RTM Nagpur University, Maharashtra, India
Department of Mechanical Engg, KDK College of Engg., Nagpur

D. LAVANYA
P.G student, Department of Mechanical Engineering, MITS, Madanapalle

G. GURU MAHESH
Department of Mechanical Engineering, MITS, Madanapalle

V. AJAY
Department of Mechanical Engineering, MITS, Madanapalle

Dr. C. YUVARAJ
Department of Mechanical Engineering, MITS, Madanapalle

Dashrath Kumar patidar
Research Scholar, Department of Mechanical Engineering, Mahakal Institute of Technology, Ujjain, M.P., India

Praveen Singh Sisodiya
Research Scholar, Department of Mechanical Engineering, Mahakal Institute of Technology, Ujjain, M.P., India

Prof. Praneeti Saxena
Associate Professor, Department of Mechanical Engineering, Mahakal Institute of Technology, Ujjain, M.P., India

Satish Kumar Patel
Student, Bachelor of Engg. Mechanical Engineering Kirodimal Institute of Technology, Raigarh C.G. Raigarh, Chhattisgarh, INDIA

Rajendra Kumar Patre
Student, Bachelor of Engg. Mechanical Engineering Kirodimal Institute of Technology, Raigarh C.G. Raigarh, Chhattisgarh, INDIA

Shankar Dayal Patel
Student, Bachelor of Engg. Mechanical Engineering Kirodimal Institute of Technology, Raigarh C.G. Raigarh, Chhattisgarh, INDIA

Prakash Kumar Sen
Faculty, Mechanical Engg. Department Kirodimal Institute of Technology, Raigarh C.G. Raigarh, Chhattisgarh, INDIA

Shailendra Kumar Bohidar
Faculty, Mechanical Engg. Department Kirodimal Institute of Technology, Raigarh C.G. Raigarh, Chhattisgarh, INDIA

SUPREET MAHADEOKAR
Research Scholar, Department of Mechanical Engineering, SRIT, Jabalpur (M.P.) India

ANSHUL CHOUDHARY
Assistant Professor, Department of Mechanical Engineering, SRIT, Jabalpur (M.P.) India

Anandkumar S Malipatil
VTU, Gulbarga

Manjunath Bandi
VTU, Gulbarga

M. Senthil kumar
Asst. Professor, Dept. of Mechanical Engg.,Sona College of Technology, Salem

Dr. S. Ragunathan
Principal, Jayalakshmi Institute of Technology, Thoppur

M. Suresh
M.E, Engineering Design, Sona College of Technology, Salem

K Nidhul
PG Student, Government College Of Engineering Thrissur

Sandeep Gujaran
PG Scholar Department of Mechanical Engineering, SKN Sinhgad Institute of Technology & Science, Lonavala, Maharashtra India

Shivaji Gholap
Assistant Prof. Department of Mechanical Engineering, Sinhgad Institute of Technology & Science, Lonavala, Maharashtra, India

Dhananjay Patel
Department Of Mechanical Engineering, Vishwakarma Institute Of Technology, Pune 411 037, Maharashtra, India

K Nidhul
PG student, Govt. Engineering College Thrissur

A S Sunil
Associate Professor, Govt. Engineering College Thrissur

V Kishore
Assistant Professor, Govt. Engineering College Kannur, kish

Monaaf D. A. Al-Falahi
Faculty of Engineering, Multimedia University, Malaysia

K Suresh Bollimpelli
ASR College of Engineering & Technology, bollimpelli

V Ravi Kumar
ASR College of Engineering & Technology, K Suresh Bollimpelli, Flat No: 104 Spurthi Towers Tanuku. PIN 534211

Nikhil S Raut
PG Student, Snehankit colony, karve nagar, Pune

Ashwin Mahindrakar
Assistant Professor, Snehankit colony, karve nagar, Pune

Niraj Raja
M.Tech, Heat Power Engineering, Department of Mechanical Engg., B.I.T. Ballarpur, Dist. Chandrapur, Maharashtra, India

Dr. M. Basavaraj
Professor, Department of Mechanical Engg, B.I.T. Ballarpur, Dist. Chandrapur, Maharashtra, India

Prof. Avinash Khanderao
Asst. Professor, Department of Mechanical Engg, B.I.T. Ballarpur, Dist. Chandrapur, Maharashtra, India

S Rathnavel
PG Scholar, Department of Aeronautical Engineering, Regional Centre of Anna University, Tirunelveli-627007, India

K Vijaya sankaran
PG Scholar, Department Aeronautical Engineering, Park College of Engineering and Technology, Coimbatore-641668, India
Vijaya sankaran K, 2/132-1 modern nagar, collectorate post, virudhunagar- 626002

Tushar S. More
PG Student, Sai Shilpa society, Bibvewadi, Pune

Sharayu U. Ratnaparkhi
Assistant Professor, Sai Shilpa society, Bibvewadi, Pune

Chaitanya G Burande
PG Student, Sai Shilpa society, Bibvewadi, Pune

Sanjaysingh Vijaysingh Patil
Assistant Professor, Sai Shilpa society, Bibvewadi, Pune

Kelkar
Mechanical engineering department Sinhgad institute of technology lonawala

S. M. Gaikwad
Mechanical engineering department Sinhgad institute of technology lonawala

Madan M. Jagtap
Saraswati College of Engineering, India

Ashir Hawaldar
Sarasvati College of Engineering, India

Niraj Jgtap
Sarasvati College of Engineering, India

Saurabh Jain
Sarasvati College of Engineering, India

Hrushikesh D. Sawant
PG-Student, Mech. Engg. Dept. SCOE, Kharghar, Navi Mumbai, Maharashtra

S. N. Teli
Associate Professor & Head, Mech. Engg. Dept. SCOE, Kharghar, Navi Mumbai, Maharashtra

L. M. Gaikwad
Asst. Professor, Mech. Engg. Dept. BVCOE, Belpada, Navi Mumbai, Maharashtra